NUREG-1918

Phenomena Identification and Ranking Table Evaluation of Chemical Effects Associated with Generic Safety Issue 191

Manuscript Completed: August 2008
Date Published: February 2009

Prepared by
R.T. Tregoning[1], J.A. Apps[2], W. Chen[3], C.H. Delegard[4], R. Litman[5],
D.D. MacDonald[6]

[1]U.S. Nuclear Regulatory Commission
Washington, DC 20555-0001

[2]Lawrence Berkeley National Laboratory
1 Cyclotron Road
Berkeley, CA 94720

[3]Dow Chemical Co.
2301 N. Brazosport Blvd.
B-1603
Freeport, TX 77541

[4]Pacific Northwest National Laboratory
P.O. Box 999
Richland, WA 99352

[5]Radiochemistry Laboratory Basics
28 Hutchinson Drive
Hampton, NH 03842

[6]Pennsylvania State University
201 Steidle Building
University Park, PA 16802

J.P. Burke, NRC Project Manager

NRC Job Code N6100

Office of Nuclear Regulatory Research

ABSTRACT

The U.S. Nuclear Regulatory Commission (NRC) issued Generic Letter 2004-02, "Potential Impact of Debris Blockage on Emergency Recirculation during Design Basis Accidents at Pressurized-Water Reactors (PWRs)," on September 13, 2004, as the primary vehicle for addressing and resolving concerns associated with Generic Safety Issue 191, "Assessment of Debris Accumulation on PWR Sump Performance." Additionally, the NRC staff developed a safety evaluation of industry-developed guidance to provide an accepted method for evaluating PWR sump performance as requested in Generic Letter 2004-02. However, the safety evaluation and the industry guidance document provide little guidance for assessing chemical effects. The licensees are to address chemical effects on a plant-specific basis.

Both the NRC and industry have sponsored research to provide additional information and develop some guidance for evaluating chemical effects. The NRC convened an external peer review panel to review the NRC-sponsored research conducted through the end of 2005 and to identify and evaluate additional chemical phenomena and issues that were either unresolved or not considered in the original NRC-sponsored research. A phenomena identification and ranking table (PIRT) exercise was conducted to support this evaluation in an attempt to fully explore the possible chemical effects that may affect emergency core cooling system performance during a hypothetical loss-of-coolant accident (LOCA).

The PIRT was not intended to provide a comprehensive set of chemical phenomena within the post-LOCA environment. Rather, these phenomena should be combined with important findings from past research and informed by ongoing research results. It is anticipated that knowledge gained by ongoing and completed research will be considered along with the PIRT recommendations to identify and resolve existing knowledge gaps so that a more accurate chemical effects evaluation can be performed.

The PIRT panel identified several significant chemical phenomena. These phenomena pertain to the underlying containment pool chemistry; radiological considerations; physical, chemical, and biological debris sources; solid species precipitation; solid species growth and transport; organics and coatings; and downstream effects. Several of these phenomena may be addressed using existing knowledge of chemical effects in combination with an assessment of their implications over the range of existing generic or plant-specific post-LOCA conditions. Other phenomena may require additional study to understand the chemical effects and their relevance before assessing their practical generic or plant-specific implications.

FOREWORD

The U.S. Nuclear Regulatory Commission (NRC) Office of Nuclear Regulatory Research has sponsored a series of research projects to inform the NRC's regulatory decisions regarding the potential for chemical products to degrade emergency core cooling system (ECCS) performance in a pressurized water reactor (PWR) following a loss-of-coolant accident (LOCA). Specifically, the NRC has used the research to support the staff evaluation of licensee responses to Generic Letter (GL) 2004-02, "Potential Impact of Debris Blockage on Emergency Recirculation during Design Basis Accidents at Pressurized-Water Reactors (PWRs)." This GL is the principal vehicle for resolving Generic Safety Issue 191 (GSI)-191, "Assessment of Debris Accumulation on PWR Sump Performance." The NRC convened an external panel to review the NRC-sponsored research conducted through the end of 2005 and to broadly identify and evaluate additional chemical phenomena and issues that were either unresolved or not considered in the original NRC-sponsored research. NUREG-1861, "Peer Review of GSI-191 Chemical Effects Research Program," issued December 2006, summarizes this review.

The NRC also conducted a phenomena identification and ranking table (PIRT) exercise between March 2006 and June 2006 to more fully evaluate additional chemical effects that may affect ECCS performance. The PIRT panelists independently ranked the significance and current knowledge associated with chemical phenomena most likely to (1) contribute to sump screen clogging, (2) affect downstream component performance, (3) impact core heat transfer, or (4) degrade structural integrity. The PIRT identified and evaluated over 100 chemical effects phenomena. These phenomena pertain to the underlying containment pool chemistry; radiological considerations; physical, chemical, and biological debris sources; solid species precipitation; solid species growth and transport; organics and coatings; and downstream effects. This report focuses on those phenomena and their possible implications that one or more PIRT panelists identified as being potentially significant. In addition, the report identifies phenomena requiring additional consideration or evaluation.

Since this PIRT was completed, the NRC staff has initially evaluated the more than 40 phenomena that at least one PIRT panelist judged highly significant in this report. The staff expects some phenomena to improve ECCS performance and others to have no effect. The staff identified approximately 16 issues potentially deleterious to ECCS performance that merited additional scoping analysis to understand their significance. NUREG/CR-6988, "Final Report—Evaluation of Chemical Effects Phenomena in Post-LOCA Coolant," summarizes this scoping analysis and has been able to resolve all but a few of these 16 phenomena.

The staff is continuing to resolve issues associated with the potentially deleterious phenomena identified in this PIRT to assess their impact on ECCS performance. As part of this assessment, the staff will consider, as appropriate, plant-specific post-LOCA environments and conditions, results from additional vendor testing conducted since the conclusion of this scoping study, and additional evaluation to characterize the significance of these remaining issues with respect to the principal factors (e.g., particulate, chemical, and fiber concentrations) and conservatisms that licensees have used to address GL 2004-02. The staff will document the disposition of these remaining issues upon completion of this assessment.

CONTENTS

ABSTRACT iii
FOREWORD v
EXECUTIVE SUMMARY ix
ACKNOWLEDGMENTS xv
ABBREVIATIONS AND ACRONYMNS xvi
1 INTRODUCTION 1
  1.1 Description of Generic Safety Issue 191 1
  1.2 Chemical Effects Background and Related Recent Research 2
    1.2.1 ICET Program 3
    1.2.2 ICET Precipitates: Physical Characteristics and Head-Loss Testing 4
    1.2.3 Thermodynamic Analytical Simulations 6
    1.2.4 Industry-Sponsored Testing and NRC-Sponsored Confirmatory Evaluation 6
    1.2.5 Coatings Evaluation 8
  1.3 Phenomena Identification and Ranking Table Motivation 8
2 PIRT OBJECTIVES AND SCOPE 11
3 PIRT APPROACH 13
  3.1 General PIRT Process 13
  3.2 Chemical Effects PIRT 15
    3.2.1 Panel Selection 15
    3.2.2 Issue Definition and Background Information 16
    3.2.3 Scenario Development 17
    3.2.4 Chemical Effects Evaluation Criteria 19
    3.2.5 PIRT Issue Development 20
    3.2.6 PIRT Issue Assessment 22
4 CHARACTERIZATION OF PIRT FINDINGS AND USE OF RESULTS 25
5 RESULTS 27
  5.1 Underlying Containment Pool Chemistry 39
    5.1.1 Effect of Boron 39
    5.1.2 Temperature and pH Evolution and Variability 40
    5.1.3 Issues Related to Post-LOCA Buffering 41
    5.1.4 Dissolved Hydrogen Concentrations 43
    5.1.5 PIRT Research Recommendations 44
  5.2 Radiological Considerations 45
    5.2.1 Effect of Radiolysis on Chemical Environment 45
    5.2.2 Effect of Radiation on the Properties of Solid Species 49
    5.2.3 PIRT Research Recommendations 49
  5.3 Physical, Chemical, and Biological Debris Sources 49
    5.3.1 Debris Generation 50
    5.3.2 Physical Debris 52
    5.3.3 Post-LOCA Chemically Induced Debris 54
    5.3.4 Biological Debris 59
    5.3.5 PIRT Research Recommendations 60
  5.4 Solid Species Precipitation 63
    5.4.1 Post-LOCA Precipitation Mechanisms 63
    5.4.2 Thermal Effects 64
    5.4.3 PIRT Research Recommendations 66
  5.5 Solid Species Growth and Debris Transport 67
    5.5.1 Particulate Growth and Agglomeration 68
    5.5.2 Settling and Deposition 69

	5.5.3 PIRT Research Recommendations	70
5.6	Coatings and Organics	71
	5.6.1 Coatings	71
	5.6.2 Organics	73
	5.6.3 PIRT Research Recommendations	75
5.7	Downstream Effects	77
	5.7.1 ECCS Pumps	77
	5.7.2 RHR Heat Exchangers	78
	5.7.3 Reactor Core	79
	5.7.4 PIRT Research Recommendations	83
6	CONCLUSIONS	85
6.1	Underlying Containment Pool Chemistry	86
6.2	Radiological Considerations	86
6.3	Physical, Chemical, and Biological Debris Sources	86
6.4	Solid Species Precipitation	87
6.5	Solid Species Growth and Debris Transport	88
6.6	Organics and Coatings	88
6.7	Downstream Effects	89
7	REFERENCES	91

APPENDICES

Appendix A:	Curriculum Vitae for PIRT Panelists	A-1
Appendix B:	Principal Physical Attributes of the LOCA Accident Sequence	B-1
Appendix C:	PIRT Issue Summary	C-1
Appendix D:	PIRT Response Tables	D-1
Appendix E:	PIRT Panelist Evaluations of T1 Phenomena	E-1
Appendix F:	PIRT Panelist Evaluations of T2 Phenomena	F-1
Appendix G:	PIRT Panelist Evaluations of T3 Phenomena	G-1
Appendix H:	PIRT Panelist Evaluations of T4 Phenomena	H-1
Appendix I:	PIRT Panelist Evaluations of T5 Phenomena	I-1

FIGURES

Figure 1: General PIRT Process	14

TABLES

Table 1:	PIRT Panelists	16
Table 2:	Selected Post-LOCA Plant Parameters	19
Table 3:	Containment Pool Buffering Agents and pH Ranges	19
Table 4:	Importance Ranking Scheme	22
Table 5:	Knowledge Assessment Ranking Scheme	22
Table 6:	Categorization of PIRT Phenomena	26
Table 7:	Summary Rankings for T1 Phenomena	28
Table 8:	Summary Rankings for T2 Phenomena	29
Table 9:	Summary Rankings for T3 Phenomena	31
Table 10:	Summary Rankings for T4 Phenomena	34
Table 11:	Summary Rankings for T5 Phenomenon	38
Table 12:	Additional Phenomena Identified Separately from the PIRT Exercise	38
Table 13:	Post-LOCA Containment Pool Temperatures	40

EXECUTIVE SUMMARY

Generic Safety Issue 191 (GSI)-191, "Assessment of Debris Accumulation on PWR Sump Performance," was established to assess the potential for debris transport and accumulation of debris generated during a loss-of-coolant accident (LOCA) to impede or degrade emergency core cooling system (ECCS) performance in operating commercial pressurized-water reactors (PWRs). The U.S. Nuclear Regulatory Commission (NRC) issued Generic Letter (GL) 2004-02, "Potential Impact of Debris Blockage on Emergency Recirculation During Design Basis Accidents at Pressurized-Water Reactors (PWRs)," on September 13, 2004, as the primary vehicle for addressing and resolving concerns associated with GSI-191. Additionally, the NRC staff developed a safety evaluation (SE) of the May 28, 2004, Nuclear Energy Institute report, "Pressurized Water Reactor Containment Sump Evaluation Methodology." This SE provides one NRC-approved method for evaluating PWR sump performance as requested in GL 2004-02. However, the SE does not provide specific guidance for evaluating chemical effects to address concerns initially raised by the Advisory Committee on Reactor Safeguards. The SE specifies that licensees address chemical effects on a plant-specific basis.

Both the NRC and industry have sponsored research to provide additional information, and develop some guidance for evaluating chemical effects. These programs collectively have demonstrated that, under certain conditions, chemical precipitates can form that may significantly affect head loss across ECCS containment sump strainer screens. Some metrics or thresholds have also been established to indicate when significant head loss increases can result.

The NRC convened an external peer review panel to both review the NRC-sponsored research conducted up through the end of 2005 and to identify and evaluate additional chemical phenomena and issues that were either unresolved, or not considered in the original NRC-sponsored research. A phenomena identification and ranking table (PIRT) exercise was conducted to support this evaluation in an attempt to fully explore the possible chemical effects that may affect ECCS performance during a hypothetical LOCA.

The PIRT process followed the general approach that has been used and refined over the last several years at the NRC. The first step was to select the panelists and provide relevant background information on post-LOCA scenario and emergency system functions and operations. The issues and objective of the PIRT were then defined, and the LOCA scenario was characterized at distinct time intervals after the initial pipe break to provide a framework for developing issues. Next, the following four technical evaluation criteria were established for ranking chemical phenomena most likely to do the following:

(1) contribute to sump screen clogging,
(2) affect downstream component performance,
(3) impact core heat transfer, or
(4) degrade structural integrity.

Brainstorming was used to identify phenomena for consideration. The panelists completed the PIRT evaluation independently to identify the significance and knowledge level associated with each phenomenon. Rankings were finalized after discussing the rationales behind the independent rankings for those issues. Despite the discussions, occasional strong dissenting opinions arose among experts or even between one expert and the rest of the group.

The PIRT results are not intended to represent a comprehensive set of chemical phenomena or issues within the post-LOCA environment. Instead, these phenomena should be considered in the context of important findings from past research (Section 1.2) and informed by research completed since the PIRT was conducted. It is anticipated that knowledge gained by related research will be considered along with the PIRT recommendations to identify and resolve existing knowledge gaps so that a more accurate review of the chemical effects evaluations in the GL 2004-02 licensee submittals can be performed.

The PIRT phenomena fall into one of four different classifications. The first class (Category I) represents those phenomena or issues that are generally known or have been demonstrated to be significant by prior research. Additional aspects associated with these issues (i.e., variables and parameters) have also been well characterized. The second class (Category II) includes phenomena or issues that either are expected to be significant by the PIRT panelists or have been demonstrated to be significant by prior research. However, their implications with respect to ECCS performance are not well known. The third category (Category III) includes phenomena that are potentially significant but are not well understood, and the ECCS performance implications are highly uncertain. The fourth category (Category IV) represents phenomena that have no engineering significance as determined by both the aggregate PIRT rankings and individual rankings and justifications. Categories II and III summarize issues where further research could be used to address remaining uncertainties.

The significant PIRT issues are grouped topically in the report. These issues pertain to underlying containment pool chemistry; radiological considerations; physical, chemical, and biological debris sources; solid species precipitation; solid species growth and transport; organics and coatings; and downstream effects. A summary of issues is provided at the end of each topical area discussion. Emphasis is placed on those issues falling into Categories II and III that may benefit from additional evaluation.

Several of these issues (Category II) can be addressed using existing knowledge associated with the chemical effect. Additional analysis may be needed only to assess the implications of these issues over the range of existing generic or plant-specific post-LOCA conditions. Other issues (Category III) may require additional study to more fully understand the chemical effects and determine if they are relevant. Then, the practical generic or plant-specific implications can be assessed. These issues are summarized below for each topical area.

Underlying Containment Pool Chemistry

- Additional investigation on the effect of concentrated boron salts within the reactor pressure vessel (RPV) could determine whether any significant corrosion of the reactor vessel internals and fuel cladding occurs in the post-LOCA period.

- Also, the effects of concentrated borate on the types, amounts, and properties of chemical precipitates that could form within the RPV should be considered.

- Plant-specific hydrogen ion concentration (pH) and temperature history should be considered in combination with various postulated debris mixtures within the buffered environment. This evaluation would provide a necessary foundation for accurately determining the dissolved species, material corrosion and dissolution rates, and precipitate formation based on solubility and kinetic considerations. Variability in important factors should also be evaluated to determine the effect of changes or uncertainty in the underlying environment.

- Additional scoping calculations are needed to assess the magnitude and relative effect of hydrogen production from a variety of sources, including the Schikorr reaction, to determine if dissolved hydrogen plays a significant role.

Radiological Considerations

- Consideration of radiological effects is warranted to first understand the magnitude of the concentrated radioactive fields in the post-LOCA environment. Then, the effects of changes in the reduction-oxidation potential on the containment pool chemistry, material corrosion rates, and chemical speciation and solubility can be evaluated.

- The likelihood of transport and accumulation of activated species at the sump screen debris bed should be assessed to determine the strength of the resulting radiation field. Localized chemical effects resulting from radionuclides trapped within the debris bed and the effect of radiation on the properties of precipitated solid species should then be considered.

Physical, Chemical, and Biological Debris Sources

- The possibility and extent of corrosion product spallation from reactor vessel walls, primary system piping and components, and fuel cladding surfaces under conditions that would occur under post-LOCA thermal and mechanical shock should be examined.

- The possibility that potential chemical precipitates could form due to retrograde solubility at the hot reactor fuel clad surface, and subsequently spall, should be investigated.

- Plant-specific debris sources, dissolved calcium concentrations, and sump environmental conditions should be considered to assess the implications of calcium carbonate formation from atmospheric carbon dioxide on both the solid species inventory and, more importantly, its effect on precipitation kinetics. However, the amount of carbon dioxide provided by the air in the containment vessel may produce only negligible calcite compared with containment pool debris generated from other sources (e.g., fiberglass insulation, concrete, ir calcium silicate insulation).

- Realistic, plant-specific containment pool environment, materials, and concentrations should be considered to identify the appropriate corrosion acceleration or inhibition factors. One acceleration factor that should be studied is the catalytic effect of possible containment materials such as copper and citrate (or potentially other hydroxy organic-acid anions, if present).

- Possible corrosion inhibition effects include concrete and other non-metallic material aging products. Specifically, phosphate, chromate, dichromate, silicate, and borate can inhibit aluminum corrosion and deposition of copper on aluminum may inhibit or accelerate corrosion. It will be important to understand plant-specific conditions in order to credit these inhibition effects.

- The effect of significant alloying additions in nuclear structural materials and debris or compositional changes in nonmetallic materials within the containment on the material dissolution rates should be considered. This evaluation should also determine whether significant dissolved concentrations of alloying or secondary elements are present that could alter chemical effects.

- Galvanic couples that may exist within the containment pool, caused by direct contact between dissimilar metals, should be identified and evaluated to determine if they alter material corrosion rates.

- The viability and potential consequences of biological growth in a post-LOCA, borated-water environment should be investigated. If significant growth is possible within 30 days, an understanding of important post-LOCA variables that may accelerate or retard the biological growth rate would be necessary to both identify locations within the ECCS where growth is possible and evaluate the associated implications.

Solid Species Precipitation

- A realistic or conservative assessment of relevant combinations of time, temperature, pH, and ionic strength should be performed to determine the types, sizes, and concentrations of solid precipitates that may form under plant-specific conditions and to determine the significance of ripening through polymerization on these precipitates.

- Precipitation studies should also consider solid species precipitation that results from the cooling of the ECCS water at the heat exchanger and heating within the RPV.

- Research could also be useful to identify whether conservative or realistic post-LOCA environmental conditions are sufficient to foster co-precipitation, and to characterize any species that form.

Solid Species Growth and Debris Transport

- Variability in important post-LOCA variables (i.e., materials, temperatures, pH, buffering systems, etc.) should also be evaluated to determine if the effect on agglomeration is significant. This study could be used to demonstrate that both chemical precipitates created in single-effects or integrated-effects testing and surrogate precipitates produced outside the testing environment are representative of materials that form in post-LOCA conditions.

- Research on transport phenomena may be necessary to validate reducing the suspended solid particulate caused by settling and deposition. Variability in the plant design, containment materials, containment flow rates and turbulence, pH buffer, and the chemical conditions present caused by various hypothetical pipe breaks are important considerations that may affect the settling rates of containment pool debris.

Organics and Coatings

- The potentially significant coating issues to address, as indicated by the PIRT panelists, are leaching from metallic coatings and the decomposition of organics within coatings by thermolysis and radiolysis under either realistic or conservative post-LOCA conditions.

- More general research in the effects of organics should be preceded by a realistic assessment of possible post-LOCA organic sources (e.g., oils, greases, electrical insulations, plastic coatings, paints, etc.) and concentrations in the containment pool. If sufficient organic sources exist, additional research would then be useful to determine the leaching behavior of these sources and to examine the ability of the organic materials to coalesce and/or bind inorganic solids as they cool within the post-LOCA containment pool and consequently affect flow through the sump strainer screen.

- The interaction between organic materials and inorganic solids may increase the buoyancy of inorganic solids such that it restricts settling but not so great as to promote flotation. Additional study would need to determine if organic buoyancy can occur within the range of post-LOCA conditions existing at PWR plants and if the effect suspends, but does not float, significant debris concentrations that would otherwise settle. The engineering impact of organic buoyancy is significantly diminished if settling of containment pool solids is not credited in a chemical effects evaluation.

- The implications of hydrothermal hydrolysis under either realistic or conservative post-LOCA conditions have not been studied. The most susceptible coatings and insulation materials could be evaluated to determine the propensity for interactions among hydrothermal hydrolysis byproducts and with other containment pool species.

Downstream Effects

- The corrosion and erosion characteristics (i.e., hardness and particle size) of chemical products and concentrations could be assessed to determine if pump seal and internal components can be adversely affected. Magnetite and other similar particles created during crud release are expected to potentially be the most detrimental.

- The pertinent chemical effects related specifically to heat exchanger performance are deposition and clogging. A conservative analysis of deposition may be sufficient to determine if adequate heat exchanger design margins remain. The precipitation kinetics could be studied to determine if solid species concentrations at these lower temperatures are less than assumed from either aqueous concentrations or solubility considerations. Evaluation of the flow conditions at the tube inlet location may also provide information to evaluate the propensity for heat-exchanger clogging.

- Evaluation of the chemical products that precipitate and subsequently adhere to the reactor fuel could be useful to determine deposit thickness and heat transfer characteristics. The propensity for spallation of chemical products from the fuel cladding could also be studied to evaluate the additional solid products that may be formed and their implications for either reactor core or sump screen clogging.

- The effect of the physical and chemical solid debris, both formed within and transported to the reactor core, on the heat capacity of the ECCS coolant water should also be considered. However, it is expected that the debris loading would have to be significant before the heat capacity is affected. Therefore, conservative engineering evaluations may be sufficient to assess the impact on the core thermal and hydraulic transfer.

- Implications of flow blockages in locations other than the reactor core inlet nozzle and localized heating effects within individual fuel elements should also be considered. It may be possible to use analysis or credit existing operator actions to demonstrate that acceptable fuel heat transfer is maintained.

- It will be important to be mindful of potential synergistic effects related to reactor core temperature increases stemming from the downstream effect phenomena when evaluating chemical precipitation within the containment pool, at the heat exchanger, or within the reactor core.

- The production of hydrogen, hydrogen peroxide, oxygen, and hydroxyl anion within the reactor core should be considered when evaluating the reduction-oxidation potential of the ECCS water and other applicable post-LOCA chemical effects.

These issues will be resolved separately by the NRC staff as part of the closure process for GSI-191. The following additional information will be considered in resolving each of the issues above:

(1) plant-specific post-LOCA environments and conditions,

(2) research findings from industry-sponsored and NRC-sponsored chemical-effect studies completed subsequent to the PIRT,

(3) chemical effects evaluation guidance developed by the NRC, and

(4) additional scoping analysis to characterize the significance of these remaining issues with respect to the principal factors (e.g., particulate, chemical, and fiber concentrations) and conservatisms that licensees have used to address GL 2004-02.

Consideration of this information was either unavailable at the time of the PIRT or outside the scope of this exercise. If consideration of items 1–4 above is insufficient to resolve the issues identified in this report, more in-depth evaluation and study may be required. The disposition of the issues identified in this report will be documented separately.

ACKNOWLEDGMENTS

The principal authors of this report gratefully acknowledge the assistance of several people who integrally contributed to both the peer review of U.S. Nuclear Regulatory Commission (NRC)-sponsored chemical effects research and the phenomena identification and ranking table (PIRT) exercise. Two peer review/PIRT meetings were held, one at Argonne National Laboratory (ANL) and one at the Center for Nuclear Waste Regulatory Analysis (CNWRA). These meetings were graciously and efficiently hosted, and the assistance of the participating support staff at each organization is appreciated. There were several presenters at each meeting who provided critical technical information to the panel. Dr. Bruce Letellier of Los Alamos National Laboratory (LANL), B.P. Jain of the NRC, John Giscoln of the Electric Power Research Institute, and Tim Andreychek of Westinghouse summarized the post-loss-of-coolant accident (LOCA) sequence and conditions. Dr. Letellier gave the presentation, "Principal Physical Attributes of the LOCA Accident Sequence" (Appendix B) which was used both as reference information and to help establish the framework for the PIRT. Each of the laboratories conducting NRC-sponsored activities also provided current research summaries at each meeting, as follows: Dr. Bill Shack and Dr. Jong-Hee Park for ANL, Dr. Bruce Letellier and Dr. Marc Klasky for LANL, and Dr. Vijay Jain and Dr. Jude McMurry for CNWRA.

Several other individuals provided technical information during the meeting to support issue formulation and evaluation. These include the aforementioned presenters and Ralph Architzel, Shanlai Lu, and Paul Klein of the NRC's Office of Nuclear Reactor Regulation. Dr. Bruce Letellier of LANL was integral in helping to formulate the PIRT framework, organize issues, facilitate brainstorming and discussion, take meeting minutes, and determine evaluation strategies. Ann Ramey Smith and Paulette Torres of the NRC's Office of Regulatory Research are also thanked for creating tabular summaries of the issues formulated during the brainstorming sessions. Paulette Torres is also appreciated for summarizing and collating the peer review findings in NUREG-1861. This is a necessary companion document.

Finally, the authors wish to thank the various NRC reviewers of this document. They include NRC staff Ching Ng, John Burke, Ervin Geiger, Paul Klein, Shanlai Lu, John Lehning, Ralph Architzel, and Matthew Yoder and former NRC staff Steven Unikewicz. Their contributions have made this a more accurate and comprehensive document.

ABBREVIATIONS AND ACRONYMNS

10 CFR 50.46	Title 10, Section 50.46 of the *Code of Federal Regulations*
ACRS	Advisory Committee for Reactor Safeguards
Ag	silver
Al	aluminum
$AlNa_{12}SiO_5$	sodium aluminosilicate
$Al(OH)_3$	aluminum hydroxide
$AlO(OH)$	aluminum oxyhydroxide
ANL	Argonne National Laboratory
$Al_2PO_4(OH)_3$	augelite
B	boron
BO_3^-	perborates
BWR	boiling-water reactor
Ca	calcium
Ca^{2+}	calcium cations
$Ca_3(PO_4)_2$	calcium phosphate
$CaCl_2$	calcium chloride
$CaCO_3$	calcium carbonate
$Ca(OH)_2$	portlandite
Cal-Sil	calcium silicate
$CH_3CO_2^-$	acetate
$C_3H_4OH(COO)_3^{3-}$	citrate
Cl	chlorine
CNWRA	Center for Nuclear Waste Regulatory Analysis
Co	cobalt
CO_2	carbon dioxide
CO_3^{2-}	carbonate
$C_2O_4^{2-}$	oxalate
Cr	chromium
CrO_4^{2-}	chromate
$Cr_2O_7^{2-}$	dichromate
C-S-H	calcium silicate hydrate
CSS	containment spray system
Cu	copper
DBA	design-basis accident
ECCS	emergency core cooling system
EDS	energy Dispersive Spectroscopy
EPRI	Electric Power Research Institute
F	Fluorine
Fe	iron
Fe_3O_4	magnetite
FeOOH	iron oxyhydroxide
$Fe(OH)_2$	iron(II) hydroxide
$Fe(OH)_3$	iron(III) hydroxide
FSAR	final safety analysis report

GL	generic letter
GSI-191	Generic Safety Issue 191
H	hydrogen
$HAsO_4^{2-}/H_2AsO_4^{1-}$	arsenate
HELB	high-energy line break
HCO_3^-	bicarbonate
HNO_3	nitric acid
H_2O_2	hydrogen peroxide
$HPO_4^{2-}/H_2PO_4^{1-}$	phosphate
ICET	integrated chemical effects testing
In	Inconel®
LANL	Los Alamos National Laboratory
LB	large break
LOCA	loss-of-coolant accident
Mg	magnesium
Mn	manganese
$Mo_7O_{24}^{6-}$	molybdate
N_2	atmospheric nitrogen
$NaBO_2$	sodium metaborate
NaOH	sodium hydroxide
Nb	niobium
NEI	Nuclear Energy Institute
Ni	nickel
NO_3^-	nitrate
NPSH	net positive suction head
NRC	Nuclear Regulatory Commission
NRR	Office of Nuclear Reactor Regulation
NUKON®	commercial fiberglass insulation
O	oxygen
OH	hydroxyl radicals
pH	hydrogen ion concentration
Pb	lead
PIRT	phenomena identification and ranking tables
PWR	pressurized-water reactor
PZC	point of zero charge
RCP	reactor coolant pump
RCS	reactor coolant system
Redox	reduction-oxidation
RHR	residual heat removal system
RMI	reflective metal insulation
RPV	reactor pressure vessel
RWST	refueling water storage tank

S	sulfur
Sb	antimony
SCC	stress corrosion cracking
SE	safety evaluation
SiO_2	silicon dioxide
SiO_4^{4-} and SiO_3^{2-}	silicates
SO_4^{2-}	sulfate
SS	stainless steel
STB	sodium tetraborate
TH	thermohydraulic
TMI	Three Mile Island
TSP	trisodium phosphate
WCAP	Westinghouse Commercial Atomic Power
Zn	zinc
ZOI	zone of influence
Zr	zirconium
ZrO_2	zirconium oxide

1 INTRODUCTION

1.1 Description of Generic Safety Issue 191

Generic Safety Issue (GSI)-191, "Assessment of Debris Accumulation on PWR Sump Performance," [1] was established to assess the potential for debris transport and accumulation of debris generated during a loss-of-coolant accident (LOCA) to impede or degrade emergency core cooling system (ECCS) performance in operating commercial pressurized-water reactors (PWRs). The ECCS is required to meet the criteria of Title 10, Section 50.46, "Acceptance Criteria for Emergency Core Cooling Systems for Light-Water Nuclear Power Reactors," of the *Code of Federal Regulations* (10 CFR 50.46) [2]. In 10 CFR 50.46(b)(5), the U.S. Nuclear Regulatory Commission (NRC) requires that licensees design the ECCS with capability for long-term cooling. After successful initiation, the ECCS must provide adequate cooling to maintain the peak cladding temperature below 2200 °F, limit cladding oxidation, limit hydrogen (H) generation, and sustain a core geometry that is amenable to cooling.

A common hypothetical scenario addressed by GSI-191 is as follows. A high-energy line break (HELB) inside PWR containment damages and dislodges materials near the break such as thermal insulation, coatings, and concrete. Actuation of the containment spray system (CSS) then washes a fraction of this debris and preexisting debris in containment (referred to as latent debris) down onto the containment floor. Simultaneously, the containment floor is gradually filling with CSS, reactor coolant, and borated water injected from the refueling water storage tank (RWST). The borated water is injected into the reactor to ensure core cooling and provide acceptable boron concentration to ensure shutdown margin that terminates fission in the core. The post-LOCA debris comingles with water from these various sources as the containment fills. At some time, depending on the size of the break, the RWST is depleted, and ECCS pump operation is changed to draw water from the containment sump and recirculate the existing debris-laden water that has accumulated within the containment floor pool.

Once recirculation starts, a fraction of containment pool debris is transported to the sump screen where it either accumulates or is ingested into the pump suction line. If debris accumulates on the sump screen, the head loss across the screen will increase. If enough debris accumulates, the concern is that it could reach a critical loading such that the net positive suction head (NPSH) is insufficient to ensure the successful operation of the ECCS and CSS pumps.

Alternatively, debris that passes through the sump screen raises additional concerns, which are collectively referred to as "downstream effects." For instance, the debris could plug or cause excessive wear of close-tolerance components or deposit on surfaces within the ECCS or CSS. Plugging or wear might cause a component to degrade to the point where its performance characteristics fall below the design-basis requirements. Also, debris blockage at flow restrictions within the ECCS could impede coolant recirculation through the reactor core, leading to inadequate core cooling. Debris could also deposit on the reactor fuel, reducing the efficiency of heat transfer to the circulating fluid.

Chemical effects could also occur during one or several stages of this hypothetical scenario. Chemical effects are broadly defined as any chemically induced phenomenon that might contribute to either head loss across the sump screen or deleterious system performance caused by downstream effects. Possible chemical effects include the formation of chemical by-products (e.g., precipitates, scale, reduction-oxidation (redox) and dissolution) caused by the interaction of containment materials with the post-LOCA environment and degradation of containment materials caused by chemical processes within the post-LOCA environment. Containment materials literally encompass all materials present in containment and notably include LOCA-generated debris, latent debris, structural materials for reactor system components (e.g., piping, reactor pressure vessel (RPV), steam generators, heat exchangers, reactor internal structures, etc.), containment structural materials, ECCS pump, valve, and heat-exchanger internal components, intact insulation and electrical materials, and reactor fuel.

On September 13, 2004, the NRC issued Generic Letter (GL) 2004-02, "Potential Impact of Debris Blockage on Emergency Recirculation during Design Basis Accidents at Pressurized-Water Reactors (PWRs)," [3] as the primary vehicle for addressing and resolving concerns associated with GSI-191. The GL requested that all PWR licensees use an NRC-approved method (1) to perform a mechanistic evaluation of the potential for post-accident debris blockage and operation with debris-laden fluids to impede or prevent the recirculation functions of the ECCS and CSS following all postulated accidents for which these recirculation functions are required and (2) to implement plant modifications or other corrective actions that the evaluation identifies as necessary to ensure system functionality.

Additionally, the NRC staff developed a safety evaluation (SE) [4] of the May 28, 2004, Nuclear Energy Institute (NEI) report, "Pressurized Water Reactor Containment Sump Evaluation Methodology," dated May 28, 2004 (see enclosures to transmittal letter in [5]). The NEI report, as modified by the SE, provides one NRC-approved method for evaluating PWR sump performance as requested in GL 2004-02. The SE specifies that nuclear plant licensees address chemical effects on a plant-specific basis. Licensees can use existing information and generic test results by providing technical justification demonstrating the applicability to their plant(s). Licensees also have the option of pursuing plant-specific testing and analysis. However, the SE does not provide a specific method that is considered acceptable for evaluating chemical effects. The staff has prepared draft evaluation guidance (see enclosures to transmittal letter in [6]) for licensees' chemical effect analyses to ensure that chemical effect issues are adequately resolved within the context of the GL 2004-02 responses.

1.2   Chemical Effects Background and Related Recent Research

The Advisory Committee for Reactor Safeguards (ACRS) initially recommended that an adequate technical basis should be developed to resolve associated sump performance issues related to chemical reactions [7]. This recommendation was based, in part, on the gelatinous material that was observed in a water sample taken from the bottom of the Three Mile Island (TMI) containment following the accident in 1979 [8]. It has not been possible to determine the direct relevance of this finding to post-LOCA chemical effects because of the atypical nature of the TMI event. A significant amount of Susquehanna River water leaked from service water system relief valves and was mixed with the reactor coolant system (RCS) water in containment. The river water in the TMI configuration was used to provide emergency cooling to service water systems [9]. An additional confounding factor was that the containment water samples were collected approximately 5 months after the event, allowing more time for solid species reactions to occur than in the 30-day post-LOCA period of most concern. In order to assess chemical effects in post-LOCA PWR environments, the NRC and the nuclear industry have sponsored a number of research programs.

1.2.1 ICET Program

An initial limited-scope study was performed by Los Alamos National Laboratory (LANL) to evaluate potential chemical effects occurring following a LOCA [10]. This study assessed the potential for chemically induced corrosion products to impede ECCS performance. In some of these tests, aluminum (Al), zinc (Zn), or iron (Fe) nitrates were added to the test water to induce chemical precipitation between the metals and other anionic species in order to assess the associated head loss. The tests showed that gel formation with a significant accompanying head loss across a fibrous bed was possible, although the test conditions and evolution of chemical species were not prototypical of a post-LOCA PWR environment.

Therefore, a more comprehensive integrated chemical effects testing (ICET) program was conducted under the joint sponsorship of the NRC and the Electric Power Research Institute (EPRI) [11]. The primary objectives for the ICET test series were to (1) determine, characterize, and quantify chemical reaction products that may develop in the containment pool under a representative post-LOCA environment and (2) determine and quantify any amorphous or gelatinous material that could be produced during the post-LOCA recirculation phase.

The ICET program consisted of five 30-day tests with each test representing a unique environment. Each environment was intended to represent a portion of the commercial PWR plants such that the entire series would broadly characterize containment pool materials and conditions applicable to the existing US PWR nuclear plants after a hypothetical LOCA. Representative types and amounts of submerged and unsubmerged materials were evaluated based on plant surveys [12]. The primary variables that changed during each test were the buffering agent (i.e., sodium hydroxide (NaOH), sodium tetraborate (STB), or trisodium phosphate (TSP)) and the insulation materials (i.e., 100 percent fiberglass or 80 percent calcium silicate (Cal-Sil) and 20 percent fiberglass).

Companion thermodynamic simulations and corrosion and leaching studies were conducted at the Center for Nuclear Waste Regulatory Analysis (CNWRA) at the Southwest Research Institute [13] to determine the need for a pressurized test loop and the efficacy of isothermal testing to evaluate long-term (i.e., several days) chemical effects in ICET. The study predicts that the high-temperature and pressure conditions existing during the initial stages of a LOCA have a negligible influence on the long-term chemical effects. Based on this work, the ICET program was conducted at atmospheric pressure and a constant temperature. A temperature of 60 °C was chosen for ICET to broadly represent the steady-state sump temperatures predicted for several plant configurations [12].

1.2.2 ICET Precipitates: Physical Characteristics and Head-Loss Testing

The ICET results indicate that (1) chemical reaction products with varied quantities, consistencies, attributes, and apparent formation mechanisms were found in each unique ICET environment; (2) containment materials (metallic, nonmetallic, and insulation debris), hydrogen ion concentration (pH), buffering agent, temperature, and time are all important variables that influence chemical product formation; and (3) changes to one important environmental variable (e.g., a pH adjusting agent or insulation material) can significantly affect the chemical products that form. Specifically, amorphous aluminum hydroxide ($Al(OH)_3$) [14] precipitates were observed upon cooling in ICET Tests 1 and 5 and similar Al-based products were observed in the Tests 1, 4, and 5 fiberglass insulation along with other particulate deposits on the outside of the insulation bundles. Calcium phosphate ($Ca_3(PO_4)_2$) was observed within the test solution within 1 hour of initiating ICET Test 3, and similar products were apparent in fiberglass insulation in Tests 2 and 3 along with other particulate deposits on the outside of the insulation bundles. Another notable observation occurred on the eighth day of ICET Test 3 when the turbine-type flow meter stopped working. Scale and precipitation deposits, which were later determined to be a mix of Cal-Sil and $Ca_3(PO_4)_2$, had accumulated on the turbine inside the flow meter, preventing the turbine from moving [11].

The ICET program provided insights and initial understanding of the solution chemistry and of the types and amounts of chemical reaction products that could form in the ECCS containment pool. However, while the ICET program demonstrated that plant-specific differences in important variables could significantly alter the observed chemical effects, it was difficult to extrapolate the results to other conditions based on the limited testing performed. Subsequently, the NRC sponsored follow-on work at LANL to more fully understand the formation of the amorphous $Al(OH)_3$ precipitate in NaOH and STB environments (ICET Tests 1, 4, and 5). Additionally, this work evaluated the behavior of Al over a broader range of conditions (i.e., pH, temperature, and solution age) that might exist throughout the PWR following a LOCA [14].

Examination of the precipitates of ICET Tests 1 and 5 revealed largely amorphous $Al(OH)_3$ with a substantial quantity of boron (B) (up to 35% of the initial dissolved B) adsorbed onto the surface. Complexation between Al and B was found to be responsible for impeding the crystallization of Al compounds. Aluminate ions ($Al(OH)_4^-$) are the only stable form of Al in an alkaline solution. The solubility is a function of solid hydroxide phases and increases with pH in alkaline solutions. The presence of some organics and inorganics can also increase the Al solubility. The solubility can also be affected by the discrete particle size in the solution [14]. Nanometer-sized particles were found in the ICET Tests 1 and 5, and transmission electron microscopy measurements revealed that colloids, with a mean radius of 65 nm, did exist at ambient temperature in these solutions. This work also found that the corrosion of Al in ICET Test 4 was inhibited by sodium silicate formed by the dissolution of Cal-Sil.

Although the ICET program demonstrated that chemical products could form under these representative post-LOCA conditions, there was no evaluation of the head-loss characteristics or downstream implications of chemical products observed during testing. Therefore, the NRC sponsored further research at Argonne National Laboratory (ANL) to evaluate the head loss associated with chemical byproducts observed during the ICET series and to understand the effect of relevant changes within the post-LOCA environment on head loss, chemical byproduct formation, and the physical characteristics of these byproducts [15]. In contrast to the focus on Al of the LANL ICET follow-on study [14], the initial ANL research focused on conditions affecting $Ca_3(PO_4)_2$ formation and settling within TSP environments (ICET Tests 2 and 3).

The ANL testing demonstrated that calcium (Ca) from representative Cal-Sil insulation loadings and with representative TSP dissolution rates caused significant pressure drop across a test bed when the amount of dissolved Ca was greater than 25 ppm (6.2×10^{-4} M). Also, significant head loss may result from the precipitation of $Ca_3(PO_4)_2$ which forms either within the containment pool and is transported to the sump screen, or from continued Cal-Sil dissolution within a sump screen debris bed. The head losses associated with pure physical debris beds of commercial fiberglass insulation (NUKON®) and Cal-Sil were generally much smaller than those that occurred across debris beds in which some of the Cal-Sil was replaced with a corresponding amount of $Ca_3(PO_4)_2$ precipitates. Debris beds formed solely from $Ca_3(PO_4)_2$ in a TSP-buffered environment caused little pressure drop as the precipitate by itself did not form a continuous debris bed across the test screen.

Testing demonstrated that the Cal-Sil dissolution rate (for the concentrations studied) is not strongly dependent on the TSP dissolution rate for realistic TSP dissolution rates. However, even for a Cal-Sil suspended-solid concentration as low as 0.5 g/l, the dissolved Ca concentration reached levels within a few hours that led to significant head loss in these tests. Settling tests of $Ca_3(PO_4)_2$ under no bulk directional flow found that at higher dissolved Ca concentrations (300 ppm (7×10^{-3} M)) the precipitates can agglomerate and settle more quickly. However, approximately one-half of the total precipitate settles more slowly than the agglomerated precipitate. At a lower dissolved Ca concentration (75 ppm (1.9×10^{-3} M)), the estimated settling velocity is 0.8 cm/min.

ANL also evaluated head loss from Al precipitates that were intended to be similar to the amorphous $Al(OH)_3$ precipitates in ICET Tests 1 and 5. Dissolved Al from aluminum nitrate was used to produce dissolved the Al concentrations which were observed in ICET more expeditiously than in a 30-day test. The testing also demonstrated that head loss in NaOH environments could be significant down to 100 ppm (4×10^{-3} M) of dissolved Al, even though visible precipitate is not apparent. Also, head loss was shown to be a strong function of the precipitate solubility limit, which varies with Al concentration, temperature and time at temperature.

These initial studies [15] also indicated that no significant head loss in STB-buffered solutions at a pH of approximately 8.3 occurs unless the dissolved Al is greater than 50 ppm (2×10^{-3} M). Post-LOCA containment pool dissolved Al concentrations higher than 50 ppm (2×10^{-3} M) are not expected in most plants. Follow-on testing was performed to characterize the long-term solubility limits and head-loss characteristics of Al precipitates in STB-buffered environments [16]. The findings suggest that a concentration of 50 ppm (2×10^{-3} M) Al can be maintained in STB-buffered solutions with a pH of 8.4 at 70–80 °F for periods of at least 20 days without the formation of significant amount of precipitate product, or head loss. In these tests, concentrations greater than 70 ppm (3×10^{-3} M) caused measurable head loss.

1.2.3 Thermodynamic Analytical Simulations

In parallel with the experimental programs at LANL and ANL, CNWRA was contracted to evaluate the feasibility of using commercially available thermodynamic simulation codes for predicting chemical species formation in a typical PWR post-LOCA containment environment [17]. Specifically, CNWRA evaluated codes initially through blind prediction of the ICET series using the existing thermodynamic speciation database associated with each code and either published corrosion or leaching rate data or rates experimentally determined by CNWRA using small-scale testing. Follow-on calibrated analysis of the ICET series was conducted by disallowing solid precipitation of species that were not observed in ICET.

Results of this study indicate that thermodynamic simulation codes can be broadly useful in assessing the potential effects of post-LOCA interaction on sump screen blockage. However, their predictive capability is often hindered by insufficient thermodynamic data for relevant phases and aqueous species in the code database and by limitations in the kinetic data for the dissolution of reactive materials in the presence of co-dissolving materials. Prediction of chemical byproduct concentrations and species by these codes is most accurate when the system is in equilibrium. But they can be quite inaccurate if equilibrium has not been achieved. However, benchmarking the results using relevant experimental data can determine if the predicted species and concentrations are representative for carefully defined conditions.

1.2.4 Industry-Sponsored Testing and NRC-Sponsored Confirmatory Evaluation

The nuclear industry has also sponsored a study to provide input to plant licensees on the type and amounts of chemical precipitates that may form post-LOCA for testing of replacement sump screens. The study evaluated leaching and corrosion data of various containment materials over a wider range of pH and temperature conditions than those tested in the ICET program [18]. The materials, pH, and temperature conditions were determined via plant surveys. The study partitioned containment materials into 15 material classes based on their chemical composition and tested 1 material from the 10 classes presumed to have the potential for causing chemical effects in the containment sump: Al, aluminum silicate, Cal-Sil, carbon steel, concrete, electrical grade glass (i.e., E-glass), amorphous silica, Interam E class insulation, mineral wool, and Zn.

Single-material, single-effect dissolution testing was conducted and the largest dissolved concentrations consisted of Al, silicon (Si), and Ca. Based on this, it was assumed that these elements are the most likely to form precipitates. Subsequent room temperature single-material (and a few dual-material) precipitation and filterability tests of materials containing these elements were conducted and precipitates from the following materials were observed for certain pH values: Al, Fiber Frax insulation, galvanized steel, fiberglass insulation, concrete, mineral wool, and Cal-Sil (when TSP buffering was used). The STB-buffered tests did not induce any additional precipitates. The three principal precipitates were assumed to be aluminum oxyhydroxide ($AlO(OH)$), sodium aluminosilicate ($AlNa_{12}SiO_5$), and $Ca_3(PO_4)_2$ based on energy dispersive spectroscopy (EDS) analysis. None of the precipitates settled readily. These test observations are generally consistent with the earlier NRC-sponsored studies discussed above [11 and 14-16].

The results from these studies are being used for the following two related applications:

(1) to develop a tool for predicting the amount of chemical products in specific plant environments, and

(2) to develop a procedure for generating chemical surrogates for use in large-scale validation testing simulating individual plant configurations.

The predictive tool is based on the experimental testing results for determining the dissolution rate of species under plant conditions. All dissolved materials that form during each time step are then assumed to precipitate instantaneously into $Ca_3(PO_4)_2$, $AlO(OH)$, and/or $AlNa_{12}SiO_5$. The particular precipitates depend on the plant materials and type of buffer. Methods and acceptance criteria are largely aimed at ensuring that agglomeration does not occur. These criteria are provided for producing surrogate precipitates that are intended to behave similarly to the above precipitates during strainer testing.

The NRC recently published a safety evaluation [19] of the Westinghouse Commercial Atomic Power (WCAP) report [18]. To support this safety evaluation, the NRC sponsored research to confirm certain aspects of the WCAP report and the intended applications of the results. Specifically, leaching and precipitation studies were conducted by CNWRA [20]. The three main objectives of these studies were (1) to examine the assumption that sample materials selected in the WCAP report [18] were sufficiently representative that their dissolution characteristics could be generalized from one product to the others in the same materials class, (2) to more definitively characterize any observed precipitates, and (3) to obtain leaching test data for concrete with a sample-to-solution ratio that was more representative of actual plant conditions than that in the WCAP report [18].

In general, the CNWRA results did not contradict the assumption that the leaching behavior of an individual material in a given materials class is representative of all materials in the class. However, not all materials were tested to conclusively validate this assumption either. Additionally, except for some of the Cal-Sil tests, the dissolved concentrations of elements in the CNWRA test leachates were similar to or lower than those measured in the WCAP report [18]. It was speculated that either differences in actual materials, age or storage conditions, or the greater agitation of samples used in the WCAP report [18] may have led to the differences in the dissolution results.

More importantly, no visible, settled precipitates were observed to 85 days after dissolution in any of the CNWRA tests at room temperature, even for experiments using the same materials and test conditions in which precipitates formed in the WCAP report [18]. Thermodynamic simulations of the CNWRA leachant concentrations often verified that the CNWRA solutions were unsaturated with respect to likely low-temperature species. Alternatively, in some cases supersaturation with solid species was predicted. However, corresponding filterability or head-loss testing was not performed to determine whether microscopic solids existed. Finally, leaching from concrete coupons was more dilute than that measured in the WCAP report [18] for a ratio of the material area to coolant volume that was conservative compared to plant survey information. No supersaturation solid phases were observed to precipitate in this testing either, nor were any of these phases predicted by thermodynamic modeling.

Research at ANL [16] was sponsored to evaluate the physical characteristics and head-loss performance of WCAP surrogate Al precipitates relative to the precipitates generated during earlier NRC-sponsored tests. The study determined that the Al ionic concentration affects the precipitate size. Higher-concentration solutions produce larger, denser precipitates that settle more rapidly. The WCAP surrogate preparation procedures do produce fine precipitates, and the addition of a small amount of precipitate equivalent to 5 ppm (2×10^{-4} M) of dissolved Al resulted in immediate, significant head loss across a fiberglass debris bed. However, the precipitates did have a different appearance than those in earlier testing [11 and 15], and the ANL study did not attempt to determine whether these surrogates are representative of precipitates produced under post-LOCA conditions.

1.2.5 Coatings Evaluation

The ICET Program did include representative quantities of inorganic Zn-coated carbon steel specimens in the test chamber [12]. These specimens were included to represent materials having Zn-rich primer applied without an additional topcoat and materials with Zn-rich primer exposed because of delamination of the topcoat [12]. No other topcoated specimens were included in ICET, primarily because choosing a "representative" coating was not possible given the variety of qualified and unqualified [21] systems that exist in containment. Limited testing [22] shows that several qualified epoxy systems are not expected to leach substantially in alkaline environments. However, this testing typically measured only chloride (Cl), fluoride (F), total halogens, and sulfate concentrations in the analysis of the leachate analysis. Additionally, these test solutions were not representative of post-LOCA borated containment pool chemistries and were not exposed to post-LOCA radiation conditions.

Argonne National Laboratory conducted a limited open-literature search [23] for information on the leaching characteristics of coatings in nuclear power plant environments. Coating supply companies were also contacted for information on the leaching characteristics of coatings in representative PWR post-LOCA environments. The survey generally found little information on leachability, leaching rate, and the potential leaching constituents of such coatings as a function of temperature, pH, and radiolytic conditions. The bulk of the qualification testing was conducted to demonstrate that the coatings do not fail under (1) design-basis accident (DBA) pressures and temperatures [21] and (2) radiation exposures expected over the plant's service life [24]. Although the ANSI standard [21] implies that coating samples should be irradiated before DBA testing, this is not an explicit requirement. Subsequent testing of qualified [25] and unqualified (i.e., original equipment manufacturer) [26] coating specimens has evaluated the impact of radiation-induced aging on coating failure under DBA conditions.

1.3 Phenomena Identification and Ranking Table Motivation

The various research programs summarized above have significantly advanced the understanding of chemical effects with the post-LOCA containment environment. There is increased knowledge of material dissolution behavior in post-LOCA environments for a variety of common insulation and metallic materials through both single-effect and integrated testing. These programs collectively have also demonstrated that, under certain conditions, chemical products can form and significantly affect head loss across fibrous debris beds. Furthermore, these programs have also provided some understanding about the variables affecting the formation of $Ca_3(PO_4)_2$ or amorphous

Al(OH)$_3$ precipitates observed during ICET. Additionally, some metrics or thresholds have been established to indicate when significant increases in head loss can result from those precipitates.

However, there are several fundamental issues that have not been addressed by these research activities. For instance, the effect of several materials that are expected to exist in containment (e.g., coatings, organics, activated species, etc.) and the influence of certain post-LOCA conditions (e.g., realistic post-LOCA containment temperature profiles, flow through sediment, galvanic effects, radiolytic effects, and the ratio of the containment air to pool volume) on chemical effects have not been addressed by ICET or other research programs. Also, ICET and other programs held important variables constant in a given test, and a complete range of plant-specific conditions for important variables has not been studied. It was also not possible to quantify either the mass of various chemical precipitates or the exact conditions leading to precipitation of these species under representative post-LOCA conditions in order to more readily assess plant-specific implications.

The general concern is that in a plant-specific environment under actual post-LOCA conditions, significantly different chemical effects than those observed and studied based on the ICET and other single-effect testing observations may occur. This concern is buttressed by the fact that the ICET Test1 NaOH environment created amorphous Al(OH)$_3$ precipitates that significantly affected head loss in fiberglass debris beds for the ICET quantities of submerged Al. However, one test variable change in ICET Test 4— the addition of substantial Cal-Sil insulation—minimized Al corrosion through silicate surface inhibition [14], and an amorphous Al(OH)$_3$ precipitate did not form. Further, the ability of chemical effects to degrade ECCS component performance downstream of the sump screen (i.e., downstream effects) was not addressed in any of these activities.

To address these concerns, the NRC convened a panel of five experts spanning a range of applicable technical disciplines. The panel had two objectives. The first objective was to review the NRC-sponsored research conducted through the end of 2005. The programs reviewed were (1) the ICET program [11] and follow-up analysis [14], (2) chemical speciation analysis [17], and (3) accelerated chemical effects head-loss testing [15]. Technical summaries of these projects can be found in Sections 1.2.1– 1.2.3. Specifically, the reviewers were asked to comment on the general quality of the research and the applicability of the results for interpreting post-LOCA chemical effects in PWRs. The review findings are summarized in [27] and individual reviews from all the reviewers are provided in the appendices.

Generally, the reviewers concluded that the ICET program was broadly representative of plant chemical conditions and did identify some important contributing materials and chemical interactions. The reviewers also indicated that the chemical speciation analysis did demonstrate the difficulties with using commercially available codes without proper calibration and consideration of the chemical species that may develop in the post-LOCA environment. A fundamental problem with such codes is that they calculate the species composition at the minimum of the Gibbs energy of the system. If the system is not at the minimum Gibbs energy, use of these codes may lead to erroneous results. The reviewers agreed that more rigorous code development would be necessary to develop an adequate assessment tool that models the kinetics associated with these environments.

The reviewers shared the view that although the large-scale chemical effects head-loss testing project demonstrated that head-loss implications from ICET-observed products can be significant, more work is needed to quantify the effects. There was also a general opinion that additional smaller scale testing should be conducted to allow more rapid evaluation of chemical phenomena that have not been considered and that the range of applicability of the current results should be extended to understand the effects of key variables such as time, temperature, and pH. This latter work would allow the results to be interpreted and applied to specific plant post-LOCA environments. Additional findings and recommendations are contained in [27].

The second objective of the peer review panel was to identify additional chemical phenomena and issues that were not considered in the original NRC-sponsored research and that formed the basis for many of the concerns identified earlier in this section. A phenomena identification and ranking table (PIRT) exercise was conducted to achieve this objective and fully explore the possible chemical effects that may affect ECCS performance during a hypothetical LOCA scenario. The remainder of this report details the objective of the PIRT exercise, the approach for selecting the experts and conducting the PIRT, and a summary of potentially significant phenomena or issues that were identified. Most of these phenomena or issues were identified during the PIRT, but there were a few unique issues raised in the individual peer reviews contained in [27] that have been incorporated as well. The intent is that these phenomena and issues will be addressed, as appropriate, to ensure that chemical effects are adequately considered in the demonstration of acceptable ECCS performance as requested in GL 2004-02 [3].

There have been several nuclear applications that have used the PIRT process. Examples include the scaling of thermohydraulic codes [28], the development of various light-water reactor accident scenarios [29, 30, and 31], the assessment of containment coatings research issues [32], fission product transport during the manufacturing, operation, and accident life-cycle phases associated with tristructural-isotropic (TRISO) fuel [33]. Most recently the NRC's Office of Nuclear Regulatory Research used the PIRT process to assess aging materials degradation in commercial PWR and boiling-water reactor (BWR) plants [34]. Additionally, several PIRT exercises were conducted to evaluate debris transport in wet and dry containments during a LOCA to identify important phenomena to address within the context of GSI-191 [35 and 36].

2 PIRT OBJECTIVES AND SCOPE

The principal objective of the PIRT exercise was to fully explore the possible chemical phenomena (e.g., solids, films, precipitates and effects), which could occur during a hypothetical LOCA scenario that may affect ECCS performance and possible reactor fuel damage caused by inadequate heat removal. Chemical effects were examined starting at the RCS pipe break, which initiates the LOCA, through the ECCS injection phase and CSS initiation, through the onset of ECCS recirculation, and up to 30 days after the accident. Time periods beyond 30 days were not considered in this analysis because the required reactor decay heat removal rate is typically low by this time, and ECCS margins are typically large. Thirty days also provides sufficient time for licensees to implement any necessary additional emergency mitigation measures to ensure adequate core cooling. Hence, this is a convenient termination point for the scenario. However, as stipulated in 10 CFR 50.46, the licensee still has the responsibility for ensuring long-term reactor core cooling beyond this timeframe.

The PIRT process identified the following three types of issues:

(1) issues that had been demonstrated during the previously research activities,
(2) remaining issues or questions related to this prior research, and
(3) entirely new issues that have not been previously addressed.

The PIRT was focused on identifying issues that could significantly affect ECCS performance by either altering sump screen clogging and head loss, affecting downstream component performance, modifying the reactor core heat transfer, or influencing structural integrity of important components. The PIRT also identified the state of knowledge associated with each issue so that more mature, or known, issues could be separated from issues that are not currently as well understood. Although the focus of the PIRT was on identifying potentially deleterious chemical effects, several advantageous, or potentially advantageous, chemical effects were also identified and evaluated.

3 PIRT APPROACH

The PIRT process followed the general approach that has been used and refined over the last several years [37]. The first step was to select the panelists and provide relevant background information. The issues and objective of the PIRT were then defined, and the LOCA scenario was defined using distinct time intervals after the initial pipe break to provide context for the issue development. Next, the evaluation criteria were established in concert with the panelists.

Once the framework was established, brainstorming among the panelists and facilitators was used to identify phenomena for consideration. Summary tables of phenomena were then created and reviewed by the experts to clarify issues in the attempt to develop common understanding on the definition and implications associated with each issue. The summary table review was also used to reach consensus among the panelists on the final phenomena to be evaluated. The panelists independently completed the PIRT evaluation to identify the significance and knowledge level associated with each phenomenon. Discussion was held after the initial individual evaluations for issues where the initial independent rankings revealed a strong difference of opinion among the experts or where just one dissenting opinion existed. The PIRT rankings were then finalized after this discussion. More details on the PIRT process used to identify and evaluate post-LOCA chemical effects on ECCS performance are provided subsequently.

3.1 General PIRT Process

The PIRT process was originally developed in the late 1980s [28] to characterize issues associated with scaling thermal-hydraulic codes and assessing uncertainty. The process has evolved over subsequent applications into a relatively mature technique [34]. A PIRT is fundamentally based on expert opinion. As such, enlisting experts with the requisite technical expertise and experience is a critical step in the process. The PIRT follows a structured process that is documented for accuracy and to enhance clarity. The PIRT often initially attempts to identify all relevant technical factors or issues within a given technical area (e.g., chemical effects). Technical factors or issues are defined as processes, phenomena, condition, characteristics, etc. that may exist within the technical area and alter the implications. The PIRT then seeks informed opinions about the significance and knowledge levels associated with the issues identified. Informed opinions are provided as both quantitative rankings and qualitative justifications that support the rankings [37].

The PIRT outcomes are meant to be fluid and capture the state of knowledge existing at a particular time. Significant issues can be validated or changed as additional insights, data, and experience are acquired [37]. A PIRT can be conducted at any point during the evaluation of a specific technical area. The most opportune time is when decisions are needed to determine the general direction of future research and analysis within the technical area.

A PIRT can be used to identify fundamental issues that should be initially addressed when a technical area is first being considered. It is expected that changes in the PIRT outcomes are more likely for these issues when either the science is immature or great uncertainty exists. In mature technical areas, a PIRT can be used to determine which, if any, issues remain in order to gain sufficient understanding before developing final solutions. Alternately, the PIRT can be conducted after scoping and other initial research activities have been completed in order to identify what future work is needed. This last alternative is most applicable to post-LOCA chemical effects. Research has clearly indicated that various chemical effects can be significant. Several potential solid species precipitates have also been identified and some significance metrics or thresholds have been established for the formation and implication of these precipitates.

The nine identified formal steps in a PIRT process [37] are summarized in the Figure 1 flowchart. While the flowchart depicts a serial progression between each of the steps, iteration among the various steps is more common. Steps 1–5 are typically defined by the sponsor or facilitator of the PIRT. The expert panel provides the information developed in Steps 6–8 in concert with a facilitator. Documentation is shown as Step 9, but in reality it occurs throughout the process to capture important decisions, definitions, and ideas that are necessary to complete the PIRT.

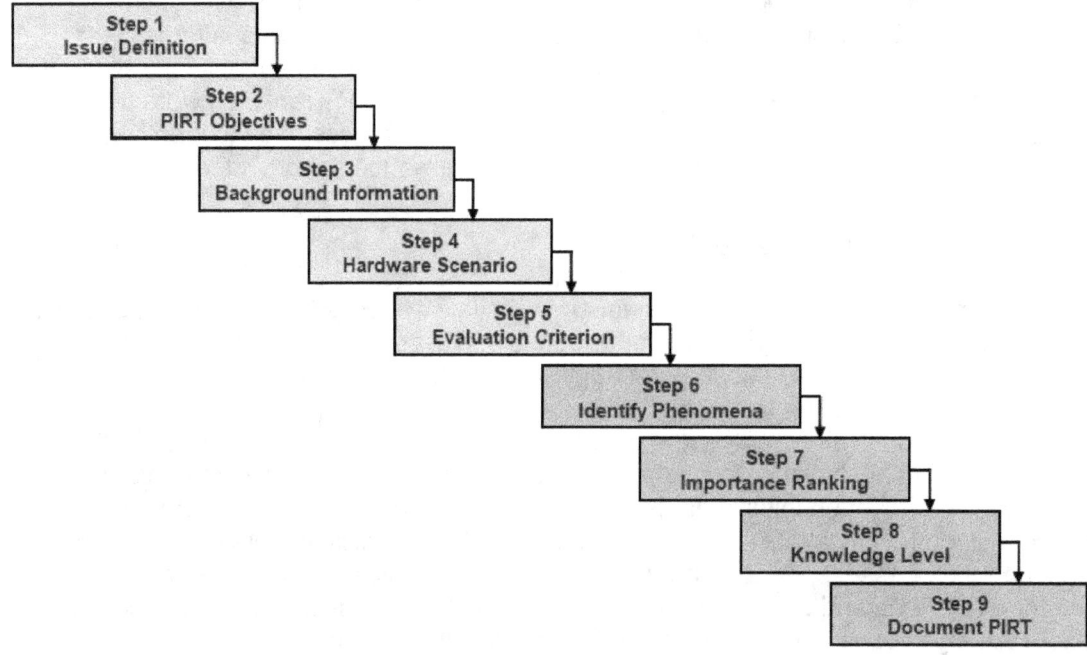

Figure 1: General PIRT Process

It is informative to briefly step through the general process as described in [37]. Issue definition (Step 1) describes the primary motivation for conducting the PIRT. It defines the technical area or most global issue that the PIRT will address. The stated PIRT objectives (Step 2) provide the specific outcome or results being sought. The focus, content, and uses of the PIRT follow from the objectives. Background information (Step 3) relates to relevant information developed before that PIRT that can be used as a basis for determining the PIRT outcomes. Some examples of background information are design information, calculations, experimental data, or presentations by the experts.

The next step (Step 4) is to identify the hardware and/or configuration that are affected by the issue and define the scenario to be evaluated. This scenario could be real or hypothetical and describes the conditions that the hardware encounters. It can often be helpful to divide the scenario into major phases [37]. The evaluation criteria (Step 5) are used to judge the relative importance of each phenomena or issue. These criteria should be directly related to implications associated with the scenario in the previous step. Ideally, the criteria are explicit, easy comprehended and measurable [37].

The phenomena or technical issues are identified next (Step 6). These issues are developed by the experts and are informed by their experience and by background information associated with the issue (Step 3). It is important that no ranking or evaluation of these phenomena occur at this step so that uncensored brainstorming can occur. Also, the issues should have precise written definitions after this step. These definitions help ensure that the experts have a consistent, common definition for evaluating the issue (Steps 7 and 8). The definitions can also be important in resolving ranking inconsistencies.

The significance or importance ranking is performed by each expert (Step 7) with respect to the evaluation criteria (Step 5). Brief written rationales are also provided to support each importance rank. Any number of ranking schemes is appropriate. In practice, consistent application of the scheme has proven to be more important than the specifics of the scheme [37]. This experience favors a simpler scheme. The knowledge assessment (Step 8) is then provided for each phenomenon or issue. Once again, written rationale is used to support the knowledge assessment, and simple assessment schemes are favored for consistency. As mentioned, PIRT documentation (Step 9) occurs during all previous steps, but it is especially important to summarize the specific PIRT approach for the technical area addressed and to document the findings.

3.2 Chemical Effects PIRT

The chemical effects PIRT followed the general PIRT framework described in the previous section. The specific application of the PIRT process for chemical effects is subsequently described.

3.2.1 Panel Selection

Expert selection was performed within the initial effort to conduct a peer review of NRC-sponsored research activities [27]. In general, the specific technical expertise common to both the peer review and PIRT was initially identified by NRC staff and contractors conducting NRC-sponsored research (Section 1.2). Technical expertise was sought in areas such as gelatinous material and amorphous product formation and characterization, analytical chemistry, metallic corrosion and leaching processes, experimental chemical testing and analysis, theoretical chemistry and speciation prediction; industrial filtration processes, and fluid treatment processes.

Once the required expertise was determined, the selection criteria were determined. The primary selection criteria were to collectively aggregate the most relevant and highest quality technical expertise while spanning all the identified technical areas. An objective was also to provide approximately equal representation among the technical areas so that the opinions consistent within, and possibly unique to, one particular discipline were less likely to dominate. Another selection criterion was organizational experience. Broad organizational representation and experience was sought so that different perspectives were available. Specifically, candidates were sought from academia, the nuclear industry, research laboratories, chemical processing industries, and filtration industries. It was determined that at least one of the experts should have nuclear plant experience in order to provide relevant, practical guidance to the remaining group members about the conditions prevalent in commercial nuclear plants.

The NRC staff, contractors, the ACRS, and nuclear industry representatives provided recommendations for suitable candidates with expertise in one or more of the identified technical areas. Staff then screened the initial pool based on the evaluation criteria. A list of 17 candidates was then developed and most were contacted to gauge availability and interest. Each candidate supplied a resume with information relevant to the evaluation criteria. The candidate resumes and independent recommendations were then used to evaluate candidates with respect to the evaluation criteria and make the final panel selections. The list and areas of technical expertise of the peer review and PIRT panelists are provided in Table 1. Curriculum vitae are provided in Appendix A.

Table 1: PIRT Panelists

Name	Affiliation	Areas of Technical Expertise
John Apps	Senior Scientist, Earth Sciences Division, Lawrence Berkeley National Laboratory	• Geochemical modeling • Gel formation and characterization • Chemical speciation modeling • Nuclear waste isolation
Wu Chen	Senior Specialist, Dow Chemical Company	• Fluid/particle separation • Industrial filtration processes
Calvin Delegard	Senior Scientist, Pacific Northwest National Laboratory	• Experimental testing and analysis • Analytical chemistry • Nuclear material safeguards
Robert Litman	Analytical Chemist, Radiochemistry Laboratory Basics	• Analytical chemistry • Metallic corrosion processes • Nuclear industry experience
Digby Macdonald	Distinguished Professor and Director of the Center for Electrochemical Science and Technology, Pennsylvania State University	• Electrochemistry and thermodynamics • Metallic corrosion processes • Experimental testing and analysis

3.2.2 Issue Definition and Background Information

The PIRT was initiated with a presentation at the PIRT kick-off meeting to articulate and develop the PIRT framework (Steps 1–5 in Figure 1). The issue (Step 1, Figure 1) was defined as follows: Operating experience, experimental testing, and analysis have demonstrated that chemical effects in post-LOCA containment environments may contribute to an inability to adequately cool the reactor core after a LOCA.

The PIRT objective (Step 2, Figure 1) was to identify chemical phenomena (e.g., solids, films, precipitates, and effects) which could lead to deleterious ECCS performance and possible reactor fuel damage caused by inadequate heat removal in a post-LOCA environment. See Section 2 for more detail. It was also indicated that the PIRT report would summarize the phenomena evaluated and indicate those significant or potentially significant phenomena that should be considered to ensure that chemical effects are appropriately addressed when evaluating ECCS performance.

Most of the background information necessary to conduct the PIRT (Step 3, Figure 1) had been previously provided to the experts as part of the general peer review. Two separate meetings were held before the PIRT to address panelist questions, conduct general discussion, and seek a common understanding of the issues. A presentation was provided during each meeting to review the PWR post-LOCA accident sequence. Current technical reports (e.g., NUREG/CR publications, test reports, and quick look reports) pertaining to the NRC-sponsored chemical effects research that has been conducted to date (summarized in Sections 1.2.1–1.2.3) were provided before each meeting. Additionally, presentations and discussions were held during the meeting with the panelists and the contractors conducting each research program. Finally, NRC personnel knowledgeable in ECCS performance and related GSI-191 issues were available at each meeting to address questions. More information on the specific background information is provided in [27].

3.2.3 Scenario Development

The post-LOCA scenario (Step 4, Figure 1) was presented at both PIRT meetings to provide the framework or basis for the exercise. The primary objectives of the presentations were to describe the principal physical attributes of the LOCA sequence and describe the function of important ECCS and related plant hardware. The scenario in the presentation at the second PIRT meeting (Appendix B) was partitioned into five distinct functional phases for subsequent assessment purposes, and a timeline was associated with each functional phase. The first phase (T1) spans the time from when the LOCA initiating event occurs at time zero (e.g., a pipe rupture) to the end of the blowdown, or pressure reduction, phase after approximately 30 seconds. Most of the LOCA containment pool debris is generated during this phase caused by the impact of the two-phase (i.e., steam plus concentrated boric acid solution), high-pressure jet emanating from the pipe break. This phase ends when the RPV and RCS pressures are largely equal to the containment pressure.

The second, or ECCS injection, phase (T2) occurs when the RWST injects borated water into the system to provide initial reactor core cooling just after the break. At the beginning of this phase, for almost all plants, the CSS activates as well to decrease the containment building pressure and temperature. This phase typically lasts from the end of blowdown (30 seconds) to approximately 20 minutes after the pipe break (for a large-break LOCA) when the RWST water has been depleted. Additional debris is created and washed into the containment pool by the CSS. Containment pool buffering is also initiated during this phase. The intended function of the buffering system is to render any radioactive iodine that has escaped from the RCS nonvolatile within the containment pool so that gaseous releases to the environment are minimized. It is important to note that the method of introduction of the buffering agent will determine whether the initial post-LOCA debris is exposed to acidic or buffered containment pool solutions. For TSP and ice condenser plants, the containment pool is initially (i.e., within the first hour or so) expected to be predominantly acidic as localized dissolution of the buffering agents occurs. Buffering in the NaOH plants should initiate more rapidly on a global scale because it is introduced through the CSS.

The third, or early ECCS recirculation, phase (T3) was defined between 20 minutes to 24 hours for the purpose of this exercise. The onset of ECCS recirculation occurs at approximately 20 minutes once the RWST water has been depleted for a large-break (LB) LOCA. The ECCS recirculation phase draws suction from the containment pool water. The water is then passed through the ECCS pumps into heat exchangers for heat rejection and back into the RPV for cooling the reactor core. The circuit is completed once water has spilled back out through the break and into the containment pool. This phase is assumed to terminate after about 24 hours because containment temperatures would have decreased from nearly saturated steam conditions (> 260 °F) at the onset of recirculation to well below the boiling point (< 180 °F) after 24 hours. Containment debris migrates to the sump screen and can accumulate on the sump screen or become ingested into the ECCS pump suction line during this phase.

The fourth and fifth phases of the event were deemed the intermediate ECCS recirculation (T4) and long-term ECCS recirculation (T5) phases, respectively. The intermediate phase was defined as 24 hours to 15 days, whereas the long-term phase was defined from 15 to 30 days. These additional phases were chosen to identify different time scales associated with corrosion or other mechanisms that may be important over the intermediate (T4 phase) and long-term (T5 phase) periods. Note that there are no new system requirements that delineate these phases. The ECCS is designed and expected to operate consistently from the onset of recirculation until circulating water cooling of the core is not required (i.e., 30 days for purposes of this exercise).

The only physical change in the system over this time period is that containment pool temperatures are continually decreasing, albeit much more slowly than decreases up to 24 hours. After 15 days, the containment temperature is expected to be less than 165 °F whereas after 30 days, the temperature is expected to be less than 145 °F. The exact temperature depends on a number of factors, including plant design, number of operating ECCS trains, and the temperature of the coolant water supplied to the ECCS heat exchangers. See Table 13 in Section 5 for more details on the post-LOCA temperature ranges.

The general plant physical layout and configuration was provided to the panel in order to provide the appropriate context for evaluating the post-LOCA scenario. The functional requirements of the ECCS components at each post-LOCA phase were described along with possible implications of component failure. Functions and characteristics of other important plant systems like the reactor vessel and containment were also discussed.

Finally, important post-LOCA parameters related to chemical effects that could affect ECCS performance were characterized. These variables include containment pool pH and buffering agent, containment materials, containment temperatures, sump inlet flow rate, containment pool height, sump configuration, and debris volumes. Representative values pertaining to each parameter were provided to the panel along with ranges to encompass plant-specific variability. The values for some of these parameters are summarized in Table 2 and Table 3, whereas more details are provided in Appendix B. These values and ranges were based on industry survey information available at the time of the PIRT. Values and ranges for some of these variables (e.g., insulation percentages and debris transport amounts) may be, or have been, altered because of plant modifications that were made to demonstrate acceptable ECCS performance as requested in GL 2004-02. The likely changes and their impact were also discussed with the panel.

Table 2: Selected Post-LOCA Plant Parameters

Parameter	Minimum Value	Mean Value	Maximum Value
Containment volume (1×10^6 ft^3)	1.3	2.6	9.9
Painted area (1×10^5 ft^2)	0.2	3.2	6.5
Level 1 coatings (1×10^5 ft^2)	0.4	3.0	6.5
Debris transported to sump screen (1×10^2 ft^3)[1]	0.01	7.1	33.8
Containment floor area (ft^2)	0.2	1.5	9.1
Maximum pool height (ft)	3.5	8.1	21
Sump flow rates (1×10^4 gpm)	0.6	1.3	2.0
Recirculation switchover times (min)	20	34	60
B concentration (1×10^3 ppm)	2.0	2.7	4.4
Reflective metal insulation in containment (%)	0	47	98
Fiberglass insulation in containment (%)	1	42	100
Mineral wool insulation in containment (%)	<10	25	60
Cal-Sil insulation in containment (%)	0	7	59

[1] Not including reflective metal insulation debris

Table 3: Containment Pool Buffering Agents and pH Ranges

	Buffer Type		
	NaOH	TSP	STB
Initial containment pH range	4–5		
Buffered containment pH range	8.5–10.5	7–8	8–9

3.2.4 Chemical Effects Evaluation Criteria

The chemical effect evaluation criteria (Step 5, Figure 1) were defined in order to judge the relative importance of each phenomenon to be assessed. Three levels of evaluation criteria were defined to establish a link between the regulatory requirements and specific technical concerns that coincide more closely with the panelists' collective expertise. The regulatory requirements contained within 10 CFR 50.46 represent the highest-level criteria. This includes requirements that the peak cladding temperature stay below 2200 °F, that maximum cladding oxidation remain below 17 percent of the total thickness, that maximum H generation remain less than 1 percent of a hypothetical amount if all metal in the cladding cylinders were to react, that the core remains coolable, and that long-term cooling is maintained for an extended period of time [2]. As mentioned, a 30-day time period was assumed in this PIRT exercise as being sufficient for demonstrating that long-term cooling is maintained.

More specific, functional level criteria were defined next. Meeting these requirements ensures that the 10 CFR 50.46 regulatory requirements are satisfied. These criteria include maintaining adequate NPSH upon consideration of debris loading at the sump strainer screens, maintaining the sufficient functioning of pumps, valves, heat exchangers, and other components downstream of sump strainer screen; and maintaining adequate heat transfer from the reactor fuel.

The lowest level criteria were next defined. These criteria are detailed technical effects commensurate with the background, experience, and expertise of the PIRT panelists. The purpose in defining these criteria was to aid the PIRT panelists in evaluating chemical phenomena (or issues) that would most likely affect the functional level and, consequently, the regulatory requirements. These criteria are effects that (1) contribute to sump screen clogging, (2) affect downstream component performance, (3) impact core heat transfer, and (4) degrade structural integrity. The panelists were specifically requested to use the preceding criteria to identify and rank the significance of post-LOCA chemical phenomena

3.2.5 PIRT Issue Development

Several underlying assumptions were used in developing the list of phenomena evaluated in the PIRT (Step 6, Figure 1). It was first assumed that effects of major fuel damage caused by failure of the ECCS to reject heat from the reactor fuel would not be considered. This assumption was made because the purpose of this PIRT is to understand how chemical effects may lead to ECCS performance failures, not how ECCS failures may lead to more severe fuel failures. This latter scenario is beyond the plants' existing design basis, but it is assessed in severe accident analysis.

It was also assumed that only degraded ECCS performance caused by chemical effects would be considered in the PIRT. Issues unrelated to chemical effects are currently addressed within the GSI-191 resolution process. An example is pump degradation caused by ingested physical debris. These issues are not related to chemical effects. However, they were discussed to provide context for assessing additional possible contributions caused by chemically induced effects. For instance, the failure to maintain the NPSH margin because of sump screen clogging by fibrous debris was not specifically evaluated by the PIRT. However, it was discussed that fibrous debris beds may be present and serve as a filter for trapping particulate and chemical byproducts. This debris may also contribute to chemical effects through reactions within the debris bed that may further alter the NPSH margin or pump performance. The panelists were asked to evaluate both the significance of any additional debris bed reactions and any enhanced chemical byproduct loading that may be filtered by a fibrous debris bed at the sump screen.

Once the assumptions governing issue identification in the PIRT were established, a brainstorming session was used to identify issues (Step 6, Figure 1) associated with each period sequentially in the post-LOCA sequence. For instance, chemical effects associated with the first phase of the LOCA scenario (T1 is 0 to 30 seconds) were discussed first. The panelists identified mechanisms, issues, or phenomena, which may occur during this time period, and which could have implications with respect to the previously defined evaluation criteria. In the T1 phase, issues such as the impingement of a high-pressure two-phase jet containing borated reactor water, elevation of containment pressure and temperature, and the wash down of debris by the high-pressure jet were identified for subsequent evaluation.

The panelist who initially identified each issue led a short discussion to clarify the definition and identify possible implications associated with that issue. The issue title, definition, and possible implications were captured on flipcharts and in electronic summary tables during the brainstorming session. This brainstorming procedure was conducted for all five post-LOCA time periods defined in Step 4 over an approximately 1-1/2-day period. After the meeting, additional descriptions and implications for phenomena were added to the summary tables based on meeting notes, flip chart content, and meeting transcripts. Also, implications that were not explicitly discussed during brainstorming were added to the summary tables for the panelists' consideration. The summary tables were then forwarded to the PIRT panelists for review.

Comments received from the panelists on the summary tables were incorporated and an updated version of the summary tables was developed. A conference call was held approximately 3 weeks after the brainstorming meeting. The purposes of this conference call were to (1) clarify and correct definitions and implications associated with the summary table phenomena, (2) identify additional issues for evaluation that panelists identified after the brainstorming meeting, (3) finalize the summary tables, and 4) provide instructions for completing the PIRT assessments. The final summary issue tables are contained in Appendix C.

Issues were not generally assessed during this brainstorming phase. However, a few issues were discarded or moved to other periods when they were inconsistent with the physical attributes associated with the time period of interest. For instance, the issue of organic radiolysis leading to carbonate (CO_3^{2-}) formation was initially identified during brainstorming for the ECCS injection phase (T2), which lasts from 30 seconds to 20 minutes. After some discussion, the panelists decided that significant CO_3^{2-} from organic radiolysis would evolve only over much longer times. The panelists then decided to move this issue to the long-term ECCS recirculation phase (T5). The updated summary tables (Appendix C) reflect the phases that the PIRT panelists determined by consensus for each phenomenon.

It should be also noted that issues raised in one phase could be active in other periods as well. Quite often, phenomena were specifically indicated only in the time period during which it first occurs. However, the panelists were instructed to assess the integrated significance of particular phenomena for issues spanning multiple phases. For instance, the fiber and particle debris mix associated with the break was identified as a phenomenon that is created in T1, yet has implications for T1–T5.

On occasion, the panel did separately identify some related phenomena in multiple time periods when the implications were distinct or the physical process warranted separate consideration during each time period. For instance, pH control was identified as a phenomenon for consideration during both the ECCS injection (T2) phase and the recirculation time (T3) phase. In the ECCS injection phase, NaOH-buffered plants have just begun to inject NaOH, whereas STB and TSP-buffered plants are just beginning to experience buffer dissolution. During recirculation, pH control through buffering is expected to be completed in all plants within about 4 hours. Therefore, the pH evolution in the containment pool is expected to start out as acidic (i.e., pH is approximately 4 to 5) at the beginning of T2 and then become basic (i.e., pH in the range of 8 to 10) by the end of T3. There are also several processes (e.g., radiolysis and hydrolysis of organic material) occurring simultaneously that may also affect the solution's pH. Because the implications associated with acidic and basic pH levels are different, the panelists decided to specifically evaluate the effect of pH in both time periods.

3.2.6 PIRT Issue Assessment

The assessment phase (Steps 7 and 8, Figure 1) used simple three-level ranking schemes to evaluate the significance or importance (Table 4) and the level of knowledge of the issues (Table 5) associated with each phenomenon evaluated by the PIRT panelists (Appendix C). It was agreed that three ranking levels provide sufficient distinction among issues without being needlessly complicated. PIRT panelists were instructed to perform importance ranking solely with respect to the four detailed evaluation criteria listed in Section 3.2.4. Table 4 provides the definitions and highlights the critical distinction between ranking levels. Issues that have a controlling, moderate, or minimal impact on evaluation criteria were to be rated high, medium, or low, respectively.

Table 4: Importance Ranking Scheme

Rank	Value	Definition
High (H)	3	Phenomenon has a **controlling** impact on one or more of the evaluation criteria
Medium (M)	2	Phenomenon has a **moderate** impact on one or more of the evaluation criteria
Low (L)	1	Phenomenon has a **minimal** impact on one or more of the evaluation criteria

The knowledge assessment in Table 5 is only slightly more complicated. The panelists were to consider both the level of understanding associated with each issue and the underlying uncertainty related to chemical effects in the post-LOCA scenario. For instance, continuing with the pH analogy, the effect of buffering on pH control may be generally well understood. However, plant-specific issues such as containment pool mixing, materials in the containment pool, and the effect of radiolysis may create large uncertainty in the actual pH during the accident at any one plant. The panelists therefore needed to weigh both contributions of general understanding and plant-specific variability or uncertainty to rank the level of knowledge. Known issues are determined to be those where engineering analysis can be performed to understand the practical ramifications. Issues that are unknown represent candidates for additional research, especially if the importance ranking is high.

Table 5: Knowledge Assessment Ranking Scheme

Rank	Meaning
Known (K)	**Sufficient** understanding to make assessment of practical ramifications; **small** uncertainty exists.
Partially known (PK)	**Partial** knowledge and understanding; **moderate** to **large** uncertainty remains.
Unknown (UK)	**Totally unknown** or very **limited** knowledge; uncertainty **cannot be characterized.**

The PIRT panelists were instructed to provide brief rationales identifying the primary reason(s) for the importance and knowledge assessment rankings provided for each phenomenon. The initial PIRT assessments were also to be completed independently. Six PIRT tables were provided to the panelists for completing the assessment (Appendix D). The first table (Table D-1, Appendix D) was created to provide a means for organizing the assessment. The specific implications associated with individual phenomena (Appendix C) were summarized, and broader implication categories were created to group the phenomena by their possible effects at critical plant locations (i.e., containment pool, sump screen, ECCS pumps, ECCS heat exchangers, and reactor core). The post-LOCA time periods were also identified so that effects associated with issues at distinct times could be identified. The table was intended to help the panelists graphically assess each issue's importance based on the number and type of implications, the relevance to critical locations, and the time periods associated with the mechanism. This table was optional, but a few of the panelists did use it to help develop their assessments.

The remaining five PIRT tables (Tables D-2–D-6) are analogous to the summary tables (Appendix C). There is a PIRT table for each time period and each row corresponds to an issue as in the summary table. Remaining columns are for panelists to indicate their importance ranking and rationale along wit their knowledge assessment ranking and rationale.

The panelists submitted their initial assessments, and the results were summarized. Numerical quantities of 3, 2, and 1 were assigned to high-, medium-, and low-ranked issues, respectively, from each panelist. Also, scores of 3, 2, and 1 were assigned to unknown, partially known, and known knowledge rankings, respectively. A few panelists chose to rank some issues at levels between those that are identified in Table 4 and Table 5 (e.g., issue importance ranked as medium/high). In these instances, an average numerical score between the two identified levels (e.g., 2.5 for a medium/high importance ranking) was assigned. Then, summary statistics were created for each issue. First, the average score, between 1 and 3, was determined for each ranking. All panelists were equally weighted in determining this average. Additionally, responses were binned to determine the number of high, medium, and low (and unknown, partially known, and known) responses for each issue. If a panelist's response fell between two identified levels (e.g., medium/high), the response was always binned with the higher importance or knowledge ranking. For example, responses that were marked medium/high were binned in the high category, and responses marked unknown to partially known were binned into the unknown category.

These summary issue rankings and the initial independent PIRT assessments provided by each panelist were distributed to all the panelists. The preliminary results where then discussed in two separate conference calls. The objectives of these calls were to (1) summarize issues that the majority of panelists ranked as being highly significant and (2) identify and discuss issues that one or two panelists ranked as highly important but were judged less significant by the remainder of the panelists. Discussion centered on identifying reasons for differences in the individual importance rankings for these issues.

The primary outcome of the group discussion was to determine which issues should be identified as significant and which could be considered less important. All significant issues are summarized in Section 5 of this report. Further discussion revealed that typically either one of two reasons occurred for divergent opinions. The most common reason was that panelists either misunderstood the issue definition or did not take into consideration some relevant background information. Panelists provided updated rankings in these cases, as appropriate, based on their new understanding of the issue (Appendices E–I). Note that the rationale was not always updated for the adjusted rankings indicated in Appendices E–I. There were only a few issues where a significant technical difference of opinion remained among more than one PIRT reviewer after further discussion. In these cases, the divergent rationale provided by the reviewers is summarized in Section 5.

There were other issues that were highly ranked by only one panelist. The high-ranking panelist for each issue determined if additional discussion of the rationale behind the high ranking was needed. Additional discussion was particularly useful if the high-ranking panelist believed that others did not appropriately consider his rationale. The alternative was that the panelist could agree with the lower rankings of the remaining PIRT members based on their technical justifications and additional clarification of the phenomenon. Typically, panelists adjusted their high rankings downward based on other panelists' justifications, as documented in Appendices E–I. A few panelists did retain their high ranking and rationale after additional discussion among the PIRT panel. In no instance was the dissenting panelist able to convince other panelists to elevate their rankings based on the dissenter's justification, although the importance of a few issues in other time periods was reaffirmed. Those issues having a single high ranking are also summarized as significant issues for consideration in Section 5. The rationale associated with both high and lower significance rankings are also summarized in this section.

4 CHARACTERIZATION OF PIRT FINDINGS AND USE OF RESULTS

Four different classifications capture the results of the PIRT evaluation, and they are summarized in Section 5. The first category (Category I) represents those phenomena or issues that are generally known, or have been demonstrated, to be significant by prior research (Section 1.2). However, some additional aspects associated with these issues (i.e., variables and parameters) still need to be considered to accurately assess relevant chemical effects in either a generic or a plant-specific analysis. An example of this type of phenomenon is the evolution of the containment pool pH caused by the addition of a buffer, radiolysis of the water, and containment pool temperature decreases during the post-LOCA scenario. The PIRT evaluation reaffirmed the great importance of pH and temperature for assessing which chemical effects will occur in the post-LOCA containment environment. Additionally, assuming radiolytic effects on pH are inconsequential (Section 5), the evolution of pH and temperature with time are generally known by the licensees for their plant-specific reactor chemistry, ECCS chemistry, and buffering system. Therefore, it is possible to address the influence of pH and temperature on post-LOCA chemical effects.

The second category (Category II) includes phenomena or issues that are either expected to be significant by the PIRT panelists or have been demonstrated to be significant by prior research (Section 1.2). However, the implications of these phenomena with respect to ECCS performance are not well known. Additional aspects of these issues (i.e., variables and parameters) should be evaluated in order to accurately assess the relevant chemical effects either generically or in a plant-specific analysis. More research or analysis could also be used to better characterize the issue by identifying significant thresholds or ranges for critical parameters and comparing those with conditions existing in plant environments.

For example, an understanding of the particular insulation materials in the containment pool after the break debris is an important consideration that will determine the type and significance of chemical effects that will occur. This mix could both alter the containment pool chemistry and the debris bed formation on the sump screen, which would lead to variability in the capture efficiency of chemical byproducts by the debris bed. The importance of this phenomenon was demonstrated in ICET [11]. Severe Al corrosion occurred during Test 1, whereas almost no corrosion was evident in Test 4. The only difference was that the copious amount of Cal-Sil that was present in ICET Test 4 increased the concentration of dissolved silica, which in turn precipitated on the Al surface and inhibited Al corrosion [14]. However, the threshold and range of dissolved silicate necessary to inhibit Al corrosion is not known (as a function of surface area), nor is there an understanding of what silicate sources in the containment (e.g., Cal-Sil and insulation) can lead to Al corrosion inhibition.

The third category (Category III) includes phenomena that are potentially significant, but are not well understood, and the ECCS performance implications are highly uncertain. These issues require more research, analysis, or evaluation within the context of specific post-LOCA plant environments to accurately consider the effects. Conservative scoping analyses could be performed initially to determine if either critical parameter ranges, bounding values, or threshold values that are necessary for the issue to be significant are representative of commercial plant conditions. If conservative analysis cannot be used to evaluate the significance, more refined research or analysis may be necessary.

For example, the effect of radiolysis on the redox potential of the containment pool chemistry has not been considered in prior research. The PIRT panelists identified this as a potentially significant consideration. Scoping calculations could be performed to conduct a first-order conservative assessment of the significance of radiolysis. If the conservative assessment cannot demonstrate that radiolysis has an insignificant effect on the redox potential, more rigorous analysis or consideration may be required.

The fourth category (Category IV) represents phenomena that are insignificant as determined by both the aggregate and individual PIRT rankings and justifications. This class includes all the issues evaluated during the PIRT but not deemed significant. An example is unsubmerged material corrosion. The PIRT panel thought that these contributions were generally insignificant compared to dissolved contributions to the containment pool chemistry from submerged materials. The Category IV phenomena are summarized in the results tables (Section 5), but are not discussed within this report.

Table 6: Categorization of PIRT Phenomena

Category	Phenomena	Implications
I	Significant	Known
II	Significant	Not well known
III	Potentially significant	Not well known
IV	Not significant	Not applicable

It is important to note that the PIRT used the prior research results (Sections 1.2.1–1.2.3) as a starting point for the exercise and did not reaffirm those findings. Instead, important research findings were discussed, and less obvious aspects that may not have been considered in the earlier research were evaluated. For example, Al corrosion was demonstrated to be important in the ICET testing [11]. The PIRT did not reassess this finding, but instead it evaluated phenomena that might alter Al corrosion rates and precipitation such as alloying, inhibition, catalytic effects, co-precipitation, etc. Therefore, the PIRT results are not intended to be a comprehensive set of chemical phenomena or issues within the post-LOCA environment. These phenomena should be combined with important findings from past research (Sections 1.2.1–1.2.3) for developing a more comprehensive list.

Additionally, since the PIRT was completed, additional research has been conducted by the nuclear industry and additional research has been sponsored by the NRC (see Sections 1.2.4–1.2.5). In some cases, this additional research has further defined critical post-LOCA variables and parameters for specific phenomena. For example, industry-sponsored testing has measured Al corrosion rates under a broader range of pH and temperature conditions than in ICET [18]. It is therefore anticipated that the knowledge gained by past and ongoing research (Section 1.2) will be considered with the PIRT recommendations to identify and resolve existing gaps so that the implications of chemical effects on ECCS performance.can be adequately understand and evaluated.

5 RESULTS

The individual PIRT rankings and rationale provided by each panelist are summarized in several appendices to this report (i.e., Appendix E–Appendix I). The first tables in each appendix replicate the issue definition and implication tables for each time period contained in Appendix C, and individual rankings were developed using the PIRT evaluations (Appendix D). Average importance and knowledge rankings were calculated as detailed in Section 3.2.6. These average rankings are summarized in Table 7–Table 11 along with the individual panelist rankings. The categories assigned to each PIRT phenomenon, as discussed in Section 4, are also included in these tables. There were also a few additional phenomena identified by individual panelists in their review of chemical effects research sponsored by the NRC [27]. Although these issues were raised after the PIRT was completed, the most significant ones are summarized in Table 12 to develop a more comprehensive list of issues for consideration.

This section will focus on discussing those issues summarized in Table 7–Table 12 that a majority of the PIRT reviewers believe are highly significant and have not been specifically addressed within activities sponsored by the NRC (Section 3.2.5). In many instances, consensus was reached on these issues and the rationale supporting the rankings will be summarized. After discussion, a few issues still had one or two panelists who believed that the issue remains highly significant even through the remainder of the panelists disagreed (Section 3.2.5). These issues and the competing rationale used to support either the high or lower rankings will also be presented.

However, in this summary, the issues are presented topically. Similar or identical issues in distinct time periods are combined under one topical entry. These topical areas do not represent a complete nor unique phenomenological description of chemical effects in the post-LOCA environment. They are merely used for grouping the significant outstanding issues. Significant issues were identified pertaining to the underlying containment pool chemistry; radiological considerations; physical, chemical, and biological debris sources; solid species precipitation; solid species growth and transport; organics and coatings; and downstream effects. This report also discusses unique phenomena within each topical area as appropriate.

At the end of each topical area discussion, a summary of issues that could benefit from additional research is provided. This summary is largely intended to capture those phenomena falling in Categories II or III (Section 4) because Category I phenomena are expected to be sufficiently mature to allow for engineering evaluation. Recall that Category II phenomena (Section 4) are generally well known and expected to be significant, but the implications with respect to either generic or plant-specific ECCS performance are not well known. Category III phenomena (Section 4) are potentially significant but are not well understood, and the ECCS performance implications are highly uncertain. The PIRT knowledge rankings and the supporting rationales are used to assign each significant phenomenon in Categories I, II, or III. Items with higher knowledge rankings (i.e., there is less known about the phenomenon) have generally been assigned to higher categories.

Table 7: Summary Rankings for T1 Phenomena

Issue	Phenomena	Average Importance (H, M, L)[1]	JA[2]	WC[2]	CD[2]	RL[2]	DM[2]	Average Knowledge (K, PK, UK)[3]	JA	WC	CD	RL	DM	Category[4]
T1-1	Crud release	2.4	L	M	H	H	H	1.6	PK	PK	PK	K	K	III
T1-2	RCS coolant conditions at break	2.4	L	M	H	H	H	1.4	K	K	PK	K	PK	II
T1-3	pH variability	2.4	L	M	H	H	H	1.4	K	K	PK	K	PK	II
T1-4	Localized B concentration in jet	1.9	L/M	M	M	L	H	1.8	PK	PK	PK	K	PK	III
T1-5	RCS fluid creates "oxidizing environment"	1.6	L	L	L	M	H	1.6	K	PK	PK	K	PK	III
T1-6	Jet impingement	2.2	H	L	H	L	H	1.8	PK	PK	PK	K	K	II
T1-7	Break proximity to organic sources	1.9	M/H	L	M	L	H	2.0	UK	PK	PK	PK	K	III
T1-8	Break proximity to secondary systems	1.2	L	M	L	L	L	2.2	UK	PK	PK	PK	PK	IV
T1-9	Debris mix particle/fiber ratio	2.6	H	H	H	M	M	1.6	PK	K	K	PK	PK	II
T1-10	Hydrogen peroxide effects	1.4	L	L	L	M	M	1.4	PK	PK	K	K	K	IV
T1-11	Nuclei formation	1.2	L	L	M	L	L	2.0	UK	PK	PK	K	PK	IV

[1] High (H), medium (M), low (L) (see Table 4 in Section 3).
[2] John Apps (JA), Wu Chen (WC), Calvin Delegard (CD), Robert Litman (RL), Digby Macdonald (DM).
[3] Known (K), partially known (PK), unknown (UK) (see Table 5 in Section 3).
[4] See Table 6 in Section 4.

Table 8: Summary Rankings for T2 Phenomena

Issue	Phenomena	Average Importance (H, M, L)	JA	WC	CD	RL	DM	Average Knowledge (K, PK, UK)	JA	WC	CD	RL	DM	Category
T2-1	Hydrogen sources within containment	1.2	L	L	L	L	M	1.6	PK	PK	PK	K	K	III
T2-2	ECCS injection of B	2.2	H	L	H	M	M	1.4	K	PK	K	PK	K	II
T2-3	Containment spray corrosion	1.6	L	M	L	M	M	1.2	K	K	K	PK	K	IV
T2-4	NaOH pH control	2.2	M	L	H	H	M	1.0	K	K	K	K	K	I
T2-5	STB pH control	1.6	M	L	M	L	M	1.0	K	K	K	K	K	I
T2-6	TSP pH control	2.2	M	L	H	H	M	1.0	K	K	K	K	K	I
T2-7	Containment spray transport	2.4	H	H	M	L	H	2.0	PK	K	K	K	UK	II
T2-8	Containment spray CO$_2$ scavenging	1.9	L/M	L	M	M	M	1.2	PK		PK	K	K	I
T2-9	Debris dissolution begins	2.2	M	L	H	M	H	1.4	K	PK	K	K	PK	II
T2-10	Carbonate concentration	1.7	L/M	L	M	M	M	1.6	PK	PK	K	K	PK	III
T2-11	Containment pool mixing	2.0	H	L	L	M	H	1.7	UK/PK	PK	K	K	PK	II
T2-12	Boric acid corrosion of exposed concrete	1.4	L	M	M	L	L	2.2	UK	PK	PK	K	UK	IV
T2-13	Fe and Ni radiological reaction	2.0	L	M	M	M	H	1.8	PK	PK	PK	K	PK	III

Issue	Phenomena	Average Importance (H, M, L)	JA	WC	CD	RL	DM	Average Knowledge (K, PK, UK)	JA	WC	CD	RL	DM	Category
T2-14	Hydrolysis	1.2	L	L	L	L	M	2.2	UK	PK	PK	K	UK	III
T2-15	Organic complexation	1.7	L/M	L	H	L	M	2.0	PK	PK	PK	UK	K	III
T2-16	Organic sequestration	1.6	L	M	M	L	M	2.2	UK	PK	PK	UK	K	IV
T2-17	Auxiliary component cooling line failure	1.7	L/M	L	M	M	M	2.2	PK	PK	PK	UK	PK	IV
T2-18	Polymerization	2.5	H	M	M	M/H	H	1.8	PK	PK	K	PK	PK	II
T2-19	Co-precipitation	1.7	L/M	M	M	L	M	1.8	PK	PK	K	PK	PK	III
T2-20	Radiolytic environment	2.0	L	L	M	H	H	1.8	PK	PK	UK	K	K	III
T2-21	Inorganic agglomeration	2.2	M/H	L	M	M/H	H	1.6	PK	PK	K	K	PK	II
T2-22	Galvanic effects	1.6	L/M	L	M	L	M	1.8	PK		UK	K	K	III
T2-23	Deposition and settling	2.3	M/H	M	L	H	H	2.2	UK	PK	UK	PK	K	II
T2-24	Organic agglomeration	2.3	L/M	H	H	L	H	2.6	UK	PK	UK	UK	PK	III

Table 9: Summary Rankings for T3 Phenomena

Issue	Phenomena	Average Importance (H, M, L)	JA	WC	CD	RL	DM	Average Knowledge (K, PK, UK)	JA	WC	CD	RL	DM	Category
T3-1	TSP pH control	1.8	L	M	H	L	M	1.2	PK	K	K	K	K	I
T3-2	NaOH pH control	2.2	H	M	H	L	M	1.0	K	K	K	K	K	I
T3-3	STB pH control	2.2	H	M	H	L	M	1.0	K	K	K	K	K	I
T3-4	NaOH injection	1.7	L/M	L	M	M	M	1.6	PK	PK	K	K	PK	IV
T3-5	Cable degradation	1.5	L		M	L	M	1.8	PK		UK	K	K	IV
T3-6	Radiolytic environment	2.3	L/M	M	H	M	H	1.8	PK	PK	UK	K	K	III
T3-7	Fiberglass leaching	2.0	M	M	M	M	M	1.0	K	K	K	K	K	I
T3-8	Secondary system contamination	1.5	L/M	L	M	L	M	1.8	UK	PK	PK	K	K	IV
T3-9	Flow-induced nucleation	1.3	L/M	L	L	M	L	2.0	UK	PK	K	UK	K	IV
T3-10	Turbulent mixing	1.5	M	M	M	L	L	2.2	UK	PK	K	UK	PK	IV
T3-11	Quiescent settling of precipitate	2.4	M	H	L	M	H	2.2	UK	K	UK	PK	PK	III
T3-12	Electrostatic scavenging	1.4	L	L	L	L	H	2.2	UK	PK	PK	UK	K	IV
T3-13	Chemically induced settling	1.7	M/H	L	L	L	H	1.8	PK	PK	PK	PK	K	III

Issue	Phenomena	Average Importance (H, M, L)	JA	WC	CD	RL	DM	Average Knowledge (K, PK, UK)	JA	WC	CD	RL	DM	Category
T3-14	Agglomeration and coagulation	2.3	M/H	H	L	M	H	1.8	PK	PK	PK	PK	K	II
T3-15	Particulate nucleation sites	1.8	M	L	L	M	H	2.2	UK	UK	K	PK	PK	I
T3-16	Additional debris bed chemical reactions	1.9	L/M	L	M	M	H	2.4	UK	UK	UK	PK	K	III
T3-17	Sump screen: high localized chemical concentrations	1.6	L	L	M	M	M	2.2	UK	PK	PK	PK	PK	IV
T3-18	Sump screen: fiberglass morphology	1.6	L	L	M	M	M	2.0	PK	PK	UK	K	PK	IV
T3-19	ECCS pump: seal abrasion	2.0	L	L	H	H	M	2.0	UK	K	UK	K	PK	III
T3-20	Heat exchanger: solid species formation	2.4	M	L	H	H	H	1.6	PK	K	UK	K	K	III
T3-21	Heat exchanger: deposition and clogging	2.1	M/H	L	H	L	H	1.6	PK	K	UK	K	K	III
T3-22	Reactor core: fuel deposition	2.1	M/H	L	L	L	H	1.6	PK	K	PK	K	K	III
T3-23	Reactor core: hydrogen increases	1.5	L	L	H	L	H	1.5	PK	K	UK	K	K	III
T3-24	Reactor core: diminished heat transfer	2.1	M/H	L	H	L	H	1.8	UK	K	UK	K	K	III
T3-25	Reactor core: blocking of flow passages	2.3	M/H	M	H	L	H	2.0	UK	K	UK	K	PK	III
T3-26	Reactor core: particulate settling	2.1	M/H	L	H	L	H	2.0	UK	K	UK	K	PK	III

Issue	Phenomena	Average Importance (H, M, L)	JA	WC	CD	RL	DM	Average Knowledge (K, PK, UK)	JA	WC	CD	RL	DM	Category
T3-27	Reactor core: precipitation	2.3	M/H	M	H	L	H	2.0	PK	PK	UK	PK	K	III
T3-28	Exposed, uncoated concrete dissolution	1.4	L	M	M	L	L	1.4	PK	PK	K	K	K	IV
T3-29	Coatings dissolution	1.6	L	L	M	M	M	2.4	PK	UK	UK	PK	PK	III
T3-30	Boric acid corrosion	1.7	L/M	L	M	M	M	1.8	PK	PK	PK	K	PK	IV
T3-31	CO_2/Carbonate radiolysis	1.0	L		L	L	L	2.0	PK		UK	PK	K	IV

Table 10: Summary Rankings for T4 Phenomena

Issue	Phenomena	Average Importance (H, M, L)	JA	WC	CD	RL	DM	Average Knowledge (K, PK, UK)	JA	WC	CD	RL	DM	Category
T4-1	Source term: unsubmerged materials	1.0	L	L	L	L	L	1.6	PK	PK	K	PK	K	IV
T4-2	Submerged source term: Pb shielding	2.0	L	H	L	L	H	1.8	PK	PK	PK	PK	K	III
T4-3	Submerged source term: Cu	1.9	L/M	M	M	M	M	2.6	UK	PK	UK	PK	UK	III
T4-4	Submerged source term: Fe	1.8	L	H	L	H	L	1.6	PK	PK	PK	K	K	III
T4-5	Submerged source term: Al, decreased concentrations	1.4	L	L	M	L	M	1.8	PK	K	PK	PK	PK	IV
T4-6	Submerged source term: Al, increased concentrations	2.2	H	H	M	L	M	1.6	PK	K	PK	PK	K	III
T4-7	Submerged source term: Fiberglass dissolution	1.6	M	L	M	M	L	1.6	K	PK	PK	PK	K	IV
T4-8	Submerged source term: Fiberglass inhibition	1.8	M	L	M	M	M	1.8	PK	PK	PK	K	PK	III
T4-9	Submerged source term: Zn passivation	1.2	L	L	M	L	L	1.8	PK	PK	PK	PK	K	IV
T4-10	Submerged source term: Zn corrosion products	1.4	L	M	M	L	L	1.8	PK	PK	PK	PK	K	IV
T4-11	Submerged source term: Zn co-precipitation	1.2	L	L	M	L	L	2.0	PK	PK	PK	PK	PK	IV
T4-12	Zinc hydroxide dissolution (zincate)	1.4	L	M	M	L	L	1.8	PK	PK	PK	PK	K	IV

Issue	Phenomena	Average Importance (H, M, L)	JA	WC	CD	RL	DM	Average Knowledge (K, PK, UK)	JA	WC	CD	RL	DM	Category
T4-13	Submerged source term: Zn-based coatings	1.4	L	M	M	L	L	2.2	PK	PK	PK	UK	PK	III
T4-14	Source Term: seal table corrosion	1.0	L	L	L	L		2.2	PK	PK	PK	UK		IV
T4-15	Submerged source term: fire barriers	1.2	L	L	M	L	L	2.8	UK	UK	UK	UK	PK	IV
T4-16	Submerged source term: organic buoyancy	1.9	M/H	L	H	L	M	2.2	PK	UK		PK	K	III
T4-17	Submerged source term: coatings	2.0	L	L	H	M	H	2.6	UK	PK	UK	PK	PK	III
T4-18	Submerged source term: RCP oil tank failure	2.2	M	M	M	M	H	2.2	UK	UK	PK	PK	PK	III
T4-19	Submerged source term: biofilm formation	1.2	L	L		M	L	2.6	UK	PK	PK	PK	PK	IV
T4-20	Submerged source term: biologically enhanced corrosion	1.2	L	L		M	L	2.4	UK	UK	UK	PK	PK	IV
T4-21	Submerged source term: biologically enhanced H$_2$ embrittlement	1.5	L	L		M	M	2.8	UK	PK	UK	UK	PK	IV
T4-22	Submerged source term: biological growth in debris beds	1.8	L	L		H	M	2.6	UK	UK	UK	PK	PK	III
T4-23	Submerged source term: electrical insulation	1.4	L	M	L	L	M	1.8	UK	UK	K	PK	K	IV
T4-24	Reactor core: fuel spalling	1.7	L/M	L	M	M	M	1.8	UK	PK	PK	K	K	III

Issue	Phenomena	Average Importance (H, M, L)						Average Knowledge (K, PK, UK)						Category
			JA	WC	CD	RL	DM		JA	WC	CD	RL	DM	
T4-25	Radiological effects: debris bed accumulation	1.9	L/M	L	M	M	H	2.4	UK	UK	UK	PK	K	III
T4-26	Radiological effects: dissolution and oxidation changes	1.6	L	L	M	L	H	2.6	UK	UK	UK	UK	K	III
T4-27	Radiological effects: radiolytic affect on biofilms	1.2	L	L		L	M	2.4	UK	PK	UK	PK	PK	IV
T4-28	Radiological effects: redox potential changes	2.1	L/M	M	M	M	H	2.0	PK	PK	UK	PK	K	III
T4-29	Radiological effects: corrosion rate changes	1.9	L/M	M	M	L	H	1.8	PK	PK	PK	PK	K	III
T4-30	Radiological effects: agglomeration	1.5	L/M	L	M	M	L	2.8	UK	UK	UK	PK	UK	IV
T4-31	CO_2/O_2 air exchange	1.4	L	L	M	M	L	1.8	PK	PK	K	K	UK	IV
T4-32	ECCS pumps: erosion/corrosion	2.4	M	M	H	M	H	1.8	PK	K	UK	PK	K	III
T4-33	ECCS pumps: seal degradation	2.2	M	M	H	M	M	2.0	UK	K	UK	PK	K	III
T4-34	Heat exchanger: secondary contaminants	1.2	L	L	L	M	L	1.6	UK	PK	K	K	K	IV
T4-35	Heat exchanger: precipitate formation	2.3	M/H	L	H	H	M	1.8	PK	K	UK	K	PK	II
T4-36	pH drop in containment pool	1.4	L	M	L	L	M	1.8	PK	PK	K	K	UK	IV

Issue	Phenomena	Average Importance (H, M, L)	JA	WC	CD	RL	DM	Average Knowledge (K, PK, UK)	JA	WC	CD	RL	DM	Category
T4-37	Reactor core: continued deposition/precipitation	2.2	M/H	L	H	L/M	H	2.0	PK	PK	UK	K	PK	III
T4-38	Reactor core: fuel deposition spalling	2.2	M	L	H	M	H	2.2	UK	UK	UK	K	K	III
T4-39	Transport phenomena: amorphous coating	1.9	L/M	H	H	L	L	2.0	PK	PK	PK	K	UK	III
T4-40	Transport phenomena: precipitation/co-precipitation	2.2	M	M	H	M	M	2.0	UK	PK	PK	K	PK	III
T4-41	Transport phenomena: metallic scouring	1.5	L/M	M	L	L	M	2.0	UK	PK	PK	PK	K	IV

Table 11: Summary Rankings for T5 Phenomenon

Issue	Phenomena	Average Importance (H, M, L)	JA	WC	CD	RL	DM	Average Knowledge (K, PK, UK)	JA	WC	CD	RL	DM	Category
T5-1	Organic radiolysis	1.5	L		M	L	M	2.2	UK		PK	PK	PK	IV

Table 12: Additional Phenomena Identified Separately from the PIRT Exercise

Issue Number	Phenomena	Description	Implications
S-1	Conversion of N_2 to HNO_3	Radiolysis converts atmospheric N_2 to HNO_3 within sump pool.	The pH in sump pool decreases because of HNO_3 production. Buffering is not sufficient to prevent reduction.
S-2	Presence of silica	Silica in the water storage systems and the RCS affects chemical product formation	1. Silica in the water storage systems and the RCS can combine with Mg, Ca, and Al to form materials with retrograde solubility. 2. Silica can have an effect on the total mass of material precipitating.
S-3	Material aging and alloying effects	1. The exposed concrete faces and dust in the containment building are aged and will be carbonated. 2. Corrosion rate data exhibits wide variability depending on the specific corrosion conditions and the alloy.	1. Numerous metallic alloys are commercially available. 2. Aging and alloying could affect dissolution/corrosion rates. 3. Aging could affect the solid species precipitates that are formed.
S-4	Advanced metallic corrosion understanding	Issues include 1. Understanding corrosion under the wide-ranging LOCA conditions (jet impingement - immersion), 2. Synergistic corrosion effects 3. The effect of hypochlorite on corrosion 4. Effects of phosphates/salts on corrosion.	Phenomena could substantially affect corrosion rates.

5.1 Underlying Containment Pool Chemistry

5.1.1 Effect of Boron

The B concentration present within the containment pool can have an effect on coolant chemistry and subsequently material corrosion rates. It also specifically plays a role in the precipitation of $Al(OH)_3$ or $AlO(OH)$ as a result of adsorption [14]. One consideration is that the B concentrations at the time of the break can be highly variable (T1-2, Table 7). Boron concentrations can range from 2000-4000 ppm (0.2–0.4 M) at the beginning of the fuel cycle to 50 ppm (5×10^{-3} M) at the end of fuel cycle. At the beginning of the fuel cycle (high B), the RCS water that impinges on materials is acidic (i.e., pH is approximately 4.5) and hot.

One PIRT panelist is concerned about the possibility that the boric acid solution can become concentrated through droplet evaporation within the high-temperature jet emanating from the pipe break to form a highly corrosive borated aerosol solution (T1-4, Table 7). These conditions were postulated by the PIRT panel to be representative of conditions existing at Davis-Besse that led to the significant erosion of the carbon-steel reactor head [38]. The variability in the B concentrations over the fuel cycle would imply that this effect could be more severe early in the fuel cycle. Although one panelist is concerned that corrosion rates could be elevated, other panelists are not concerned because of the limited area affected by the jet impaction and because the affected materials will largely be unsubmerged above the containment pool water level. This effect is also expected to be transient and will diminish as the reactor pressure equilibrates (typically within 1 minute) and as the temperature decreases below the boiling point. Therefore, the other panelists believe that longer-term chemical effects will dominate.

The initial differences in RCS B concentration during the fuel cycle become less distinct as ECCS injection starts within seconds after the break. The ECCS injects borated water (2800 ppm (0.26 M) of B) into the RCS to cool the reactor and ensure that the fission process is terminated. If the break occurs at the end of the fuel cycle, B concentrations are still expected to be greater than 1500 ppm (0.14 M) after the end of the injection phase (approximately 20 minutes). This high B concentration results in an initial containment pool pH of 4.5 to 5.5, depending on the temperature.

The effects of B on the containment pool pH (T1-3, Table 7) can be readily determined if the B levels are known as a function of time. The initial acidic conditions will accelerate corrosion of many submerged metallic components over the first several minutes of the accident and the effects of both high and low B should be considered to accurately account for differences in the dissolved concentrations of products resulting from corrosion and/or leaching.

However, as containment pool buffering and ECCS B-injection continues, the B levels will become more uniform, and the containment pool pH will become basic. These conditions should be established within the first 20 minutes after the break, and the pH should be stabilized by the buffering within 1 to 4 hours after the break [12]. The exact time required for stabilization is a function of the buffering system and the rate of mixing within the containment pool. Therefore, differences in initial corrosion products species concentrations caused by variable B early in the break should become less important as the accident sequence progresses.

The PIRT panelists identified another mechanism for creating concentrated boric acid within the RPV through borated water evaporation in contact with hot reactor fuel (T3-27, Table 9). Post-LOCA concentration of boric acid within the reactor core caused by water evaporation is a well-known phenomenon. It is less clear whether it would be sufficient locally to affect corrosion of the fuel cladding, vessel cladding, or other internal components. The boric acid may facilitate dissolution of iron (Fe) and nickel (Ni) from reactor internal components and zirconium (Zr) from the fuel cladding. These cations could subsequently precipitate as fine hydroxides at lower temperatures within the containment pool or at the heat exchanger (Section 5.4.2). The other more subtle effect of continued boric acid concentration in the reactor would be to diminish the B concentration in the containment pool.

The ICET experiments [14] also demonstrated that B can play a large role in controlling precipitate of the Al corrosion products (T2-2, Table 8). Boron apparently caused the amorphous $Al(OH)_3$ precipitate in ICET Test 1 to remain amorphous for much longer times than without B [14]. An item for further consideration is to understand how realistic B concentrations (considering all effects identified in this section) affect precipitation of other chemical species that may exist in the post-LOCA environment. Thermodynamic data for B complexes in solution are insufficient and many B complexes are inadequately characterized for adequate modeling of B chemistry in cooling water following a LOCA.

5.1.2 Temperature and pH Evolution and Variability

A related phenomenon is the temperature and pH evolution that occurs within the containment pool during the post-LOCA scenario (T1-3, Table 7). Simulated temperatures that summarize the results from approximately six distinct analyses are provided in Table 13 at various times after a hypothetical LB LOCA [12 and 39]. Temperature ranges are provided because the actual temperature/time profile is a function of many variables, including the plant type, the reactor power, the ECCS design, the availability of ECCS trains, the break size, the break location, and the heat exchanger inlet temperature. These ranges are therefore representative of a post-LOCA scenario, but they do not necessary apply to any plant-specific pipe break scenario.

Table 13: Post-LOCA Containment Pool Temperatures

	Post-LOCA Time				
	< 10 seconds	1 hour	12 hours	1 day	15 days
Temperature range (°F)	190–270	150–255	135–210	125–180	115–165

Although the containment pool temperature often peaks several minutes after the break, it is instructive to note that the peak temperature is not reached in certain simulations until approximately 1 hour after the LOCA has occurred. Additionally, the containment pool temperatures in these simulations remain significantly above ambient conditions for several weeks. Simulations conducted out to 30 days indicate that little additional temperature drop (approximately 5 °F) may occur after 15 days.

Although the temperature is highest relatively soon after the break, the minimum containment pool pH occurs at the beginning of the post-LOCA period. As discussed in Section 5.1.1, the high initial B concentration (caused by ECCS injection and/or breaks at the beginning of the fuel cycle) results in a pH of 4.5 to 5.5. The pH then rises with time as buffering agents are added. Depending on the pH buffering system, the final steady-state containment pool pH can vary between 7 and 11 and is established between 1 to 4 hours after the LOCA [12]. The time necessary to reach the uniform, steady-state pH is a function of the rate at which the buffer is introduced and the time required for it to become well mixed within the containment pool.

The relationship between the containment pool temperature and pH is critical for understanding chemical effects. The initial acidic pH range enhances corrosion and/or dissolution of certain materials (e.g., Zn, Cal-Sil), whereas other materials (e.g., Al and NUKON®) are more susceptible within the basic pH range. Significant material corrosion or dissolution may alter the pH [15] and affect the dissolution rate of other materials. Material dissolution and/or corrosion are also generally accelerated at higher temperatures. Additionally, the solubililily of solid phases can depend strongly on temperature. Lower temperatures promote the precipitation of chemical species with prograde solubilities (i.e., those that decrease with temperature such as $Al(OH)_3$), whereas higher temperatures favor precipitation of chemical species having retrograde solubility (e.g., calcium carbonates ($CaCO_3$) and $AlNa_{12}SiO_5$). The extent of prograde to retrograde solubility depends also on the composition of the aqueous phase.

Knowledge of how the pH and temperature evolve as a function of time for different post-LOCA debris constituent scenarios is necessary to understand the interdependency among pH, temperature, and the post-LOCA debris constituents (Section 5.3) for plant-specific conditions. Evaluating a single pH and temperature history is not likely sufficient, and determining credible ranges resulting from different LOCA scenarios allows for a more thorough evaluation of likely chemical effects. Once these ranges are determined, corrosion and dissolution effects and possible subsequent precipitate formation can be more accurately assessed to determine the impact of chemical effects.

5.1.3 Issues Related to Post-LOCA Buffering

It is instructive to describe the various buffering systems because these bear directly on both the pH evolution (Section 5.1.2) and the types of chemical byproducts that may precipitate. Commercial plants primarily employ these systems to sequester radioactive iodine in the aqueous liquid phase using one of the following four methods:

(1) NaOH injection through containment sprays (T2-4, Table 8; T3-2, Table 9)

(2) TSP dissolution from containment floor baskets (T2-6, Table 8; T3-1, Table 9)

(3) for ice condenser plants, STB release as ice melts to provide containment cooling (T2-5, Table 8; T3-3, Table 9)

(4) STB use as a powder from containment floor baskets for plants without ice condensers

As mentioned, the containment pool pH increases as the LOCA continues because of the progressive addition of chemical buffers. The pH values after buffering will stabilize once the buffering addition is completed and the buffer becomes well mixed within the containment pool.

For plants using NaOH buffering, the NaOH is injected into the initial containment spray. The pH of the spray is initially high (the pH is greater than 9.5 and up to 10.5) during the approximately 20-minute NaOH injection phase. The final containment pool pH is alkaline with an expected pH range of 8.5-10.5[1]. This pH is conservatively assumed to be established within 4 hours, which coincides with the time that the coolant is recirculated once through the ECCS. However, in many plants, the pH is expected to stabilize within 1 hour because the NaOH is well dispersed initially by the sprays.

Buffering in other plants occurs by dissolution of TSP from containment floor baskets. The dissolution begins once the containment pool water reaches the bottom of the basket. This time is variable depending on the plant configuration, but the onset of dissolution is expected within 20 minutes. As dissolution begins, initially high pH is present near the baskets. The relatively quiescent containment pool likely precludes significant mixing until ECCS recirculation begins (T2-11, Table 8). Complete mixing is expected within 4 hours after the coolant is recirculated once through the ECCS. The final pool pH range for plants using a TSP buffer is approximately 7–8.

Ice condenser plants are designed to release STB with the ice melt from elevated containment baskets during a LOCA. For a LB LOCA, as the ice begins to melt but ECCS recirculation has not yet begun, the pH directly under the baskets may be higher (approximately 9) than the bulk of the containment pool. However, after recirculation begins, mixing begins to equalize the STB concentration. The containment pool pH reaches steady state (i.e., when the pH is between 8–9) after the ice bed has completely melted (generally within 1 hour for a LB LOCA), and the STB has been well mixed throughout the pool. Steady-state conditions should be established in the containment pool within a few hours after the LB LOCA [12].

The concerns raised by the panelists regarding chemical effect vary for each of these buffering approaches. The NaOH buffering results in the most alkaline environment of all the buffering agents. One panelist is concerned about ionic species contributions resulting from corrosion of unsubmerged materials by the high pH spray within about 4 hours of the accident (T2-3, T2-4; Table 8). This panelist believes that the corrosion will significantly increase the containment pool dissolved metal content because of the large unsubmerged material surface area. Other panelists believe that the time period associated with unsubmerged corrosion is too short to provide significant contributions. The ICET observations indicated that predominant metallic contributions at high pH were caused by submerged Al components and that the unsubmerged contributions were insignificant [11].

A more universal concern with NaOH is for longer time periods, more than 1 day after the LOCA (T3-2, Table 9). During this time, the containment pool will be well mixed, and the high pH may promote corrosion of metals like Al and the leaching of silica and other constituents from insulating materials such as NUKON® and Cal-Sil. Corrosion of Al, the leaching of silica, and the subsequent production of $Al(OH)_3$ precipitates were

---

[1] Note that this upper pH value is less than the value reported in the final safety analysis report (FSAR) for some plants. The higher FSAR estimates used by some plants apparently did not account for the buffering capacity of the boric acid.

observed during ICET Test 1 [11]. Interestingly, much less silica leached from NUKON® than would be predicted from single-material testing. This has been attributed to "passivation" of the NUKON® by Al [17] (See Section 1.2 for more details). Although the panelists agreed that these significant observations should be considered when evaluating chemical effects, they did not identify additional phenomena related to NaOH buffering that require further consideration.

Concerns related to TSP buffering stem from the observations in the ICET Test 3 that large amounts of $Ca_3(PO_4)_2$ precipitated within the first hour upon introducing TSP (T2-6, Table 8) [11]. The precipitation of $Ca_3(PO_4)_2$ has also been shown to lead to significant head loss under certain conditions [15]. It was demonstrated in ICET Test 3 that the phosphate was depleted within 1 day or less because of the presence of excess dissolved Ca concentrations in solution [15]. This rapid depletion may not occur in TSP-buffered containment pools with smaller dissolved Ca concentrations. However, as long as sufficient dissolved Ca is present such that the reaction is phosphate limited, it is expected that $Ca_3(PO_4)_2$ concentrations similar to ICET Test 3 will be precipitated during the post-LOCA period.

Concentrations of dissolved Ca below the point where the reaction become Ca-limited, will proportionately decrease the $Ca_3(PO_4)_2$ concentration. However, for these lower dissolved Ca concentrations, other solid species may be formed between available metal cations and phosphate anions. As demonstrated by the ICET experiments, the time period associated with these phosphate reactions can be short and may proceed to completion within 1 day depending on specific cation and anion concentrations (See Section 1.2 for more details). The PIRT panelists identified no additional concerns with TSP buffering beyond the observations from the ICET and chemical effects head-loss testing.

Buffering with STB raised the least concern among the PIRT panelists (T2-5, Table 8; T3-3, Table 9). The spray and containment pool chemistry is more neutral (the pH is less than 8.5) in plants using STB buffering than for those using NaOH buffering. Therefore, corrosion of some materials is decreased. Although corrosion is diminished, precipitate formation can still occur, especially upon cooling. Precipitation was observed in ICET Test 5 [11], but it occurred later in the test (several days after the test initiated), took longer to form upon cooling (1 day or more), and was generally less concentrated when compared to the ICET Test 1 results [11]. Additionally, the STB *does not* form insoluble materials with dissolved ions (like Ca) as does TSP [40]. Therefore, STB may be the most benign buffering system.

5.1.4 Dissolved Hydrogen Concentrations

At least one PIRT member strongly believed that an accurate accounting of dissolved H_2 concentrations is important for determining the underlying ECCS water chemistry (T2-1, Table 8). His contention is that the reduction /oxidation (redox) potential is a function of the amount of dissolved H. Hydrogen sources within the containment include the RCS inventory, the corrosion of metallic materials including the reactor fuel cladding (Section 5.7.3.3), and the Schikorr reaction.

The possible generation of H from the Schikorr reaction

$$3Fe(OH)_2 \rightarrow Fe_3O_4 + H_2 + 2H_2O$$

and related chemical transformations might contribute significantly to the H inventory in containment immediately after a LOCA. Additional H in the containment pool caused by this reaction is beneficial in one sense because it may decrease the redox potential and inhibit corrosion. This is the reason H is added to PWR and BWR primary coolant circuits. Another positive attribute of the Schikorr reaction is that it decreases gelatinous iron(II) hydroxide ($Fe(OH)_2$), which may react with other components (e.g., $Si(OH)_4$) to produce mixed oxides/oxyhydroxides and mixed hydroxides that may be more detrimental than $Fe(OH)_2$. One possible downside of this reaction is that magnetite (Fe_3O_4) is produced. However, Fe_3O_4 may be a less deleterious chemical effect than $Fe(OH)_2$. On balance, this panelist appears to stipulate that this is a beneficial chemical effect assuming that H concentrations do not lead to deflagration or other nonchemical effects.

Other PIRT panelists, however, stipulated that the H concentration is small initially and that the dissolved H additions from metallic corrosion will decrease with time. Additionally, the following points were provided to argue that H would not be an effective reducing agent in this environment:

- H is not a very effective reducing agent in the absence of a catalyst (such as platinum metal),

- H partitioning into the containment building atmosphere renders it largely inert, and

- Containment buildings have H recombiners that start at approximately 2% H in the containment atmosphere. These low levels result in negligible dissolved H to affect the redox potential.

Most panelists indicated that estimating these effects should be straightforward.

5.1.5 PIRT Research Recommendations

Several issues within this section require further consideration to ensure that chemical effects are appropriately addressed. Some additional investigation on the effect of concentrated B within the RPV is warranted. The B concentration could be estimated through scoping calculations. Then, literature surveys may be used to determine if these concentrations significantly affect the corrosion rates for reactor vessel internals and fuel cladding. Additionally, existing research may indicate whether these concentrations alter the types, amounts, and properties of chemical precipitates that could form within the RPV. The important role of B in controlling the formation of $Al(OH)_3$ precipitates was previously demonstrated in the ICET experiments.

As discussed, it is vital to consider each plant-specific pH and temperature history in combination with various postulated debris mixtures within the buffered environment. This evaluation is necessary to accurately determine the dissolved species, material corrosion and/or dissolutions rates, and formation of precipitates based on solubility and kinetic considerations. Variability in the pH and temperature history should also be evaluated. Unfortunately, this assessment does not lend itself well to generic considerations because of expected differences in the post-LOCA coolant water conditions among the U.S. PWRs. It will therefore be necessary to evaluate conditions existing at each plant.

Additional scoping calculations could be used to assess the magnitude and relative effect of H production from a variety of sources, including the Schikorr reaction. The objective of this analysis is to determine if dissolved H_2 plays a significant role in the containment pool chemistry or if these effects can be neglected.

5.2 Radiological Considerations

Radiological considerations fostered much discussion and a general agreement among the PIRT panelists that these effects have not been sufficiently addressed to understand their implications on subsequent post-LOCA chemical effects. Issues fall within the following two categories: (1) the effect of radiolysis on the post-LOCA cooling water chemical environment and (2) the effect of radiation on the physical and chemical properties of solid species that could form in this environment.

There are expected to be two significant radiation sources. The reactor fuel itself is the largest source. Although active fission reaction ceases after a LOCA, gamma radiation, caused by the radioactive decay in the fuel, is still in the 1×10^6 rad/h. The other prominent source of gamma radiation is from activated debris and chemical species. Neutron-activated corrosion products and deposits form on the fuel and reactor coolant internal components. A percentage of these deposits will spall from the surface as a result of the post-LOCA pressure/temperature blowdown transient and will be released into the cooling water.

Based on the energy released, some of the more important activation isotopes that may be expected in the post-LOCA cooling water include sodium-24 (^{24}Na), beryllium-7 (^{7}Be), chromium-51 (^{51}Cr), cobalt-58 and cobalt-60 (^{58,60}Co), manganese-54 and manganese-56 (^{54,56}Mn), ^{95,97}Zr, ^{55,59}Fe, ^{59,63}Ni, niobium-95 and niobium-97 (^{95,97}Nb), antimony-125 (^{125}Sb), ^{65}Zn, and silver-110m (^{110m}Ag). The species and concentrations present will depend on plant-specific coolant chemistry, buffering agents, and the containment materials present (e.g., insulation, structural materials and cabling). The radiation fields associated with these species are likely to be significant compared to the radiation from intact fuel if species become concentrated at the sump screen. At least one PIRT panelist estimates that radiation fields on the order of 10^3 1xrad/h would be present in the sump screen. This estimate is based on the dose measured on reactor coolant resin beds and filters following a normal shutdown where fission and corrosion products are concentrated within the beds.

5.2.1 Effect of Radiolysis on Chemical Environment

The more general concern associated with activation product decay from the reactor fuel and potentially spalled activated species (e.g., Ni, Fe, Sb, Mn, and Ag) is related to the radiolysis of the post-LOCA cooling water and dissolved constituents (T2-20, Table 8; T3-6, Table 9; T4-28, Table 10). Radiolysis is the dissociation of molecules by radiation. It is the cleavage of one or several chemical bonds resulting from exposure to high-energy flux. In the post-LOCA environment, water will form H, oxygen (O), hydrogen peroxide (H_2O_2), hydroxyl radicals (OH•), and a number of other, minor products caused by radiolysis. However, the O and H gaseous radiolysis products will be quickly removed from solution by heat and agitation leaving only H_2O_2 and OH• in solution.

Both of these species are strong oxidizing agents (i.e., they have a high tendency to accept electrons from other species).

Radiolysis can subsequently modify the redox potential of the water depending on the relative amounts of H, O, and H_2O_2 produced. The redox potential is the tendency of the medium, via the electroactive chemical species present, to acquire or lose electrons and thereby be reduced or oxidized, respectively. The H_2O_2 and O concentrations can build in the solution to potentially yield a more oxidizing environment than deaerated or even oxygenated water. The redox potential is a property of the environment by virtue of the ensemble of species contained within it and is not the sole property of any given species. Therefore, it is most readily calculated using the Mixed Potential Model [41] and corresponds to the situation where no oxidation of any substrate occurs. Experimentally, this situation is most closely achieved when using a platinum or gold indicator electrode.

As the redox potential becomes more positive, or oxidizing, an oxidizing specie's affinity for electrons increases. For metallic materials, changes in the redox potential can substantially alter the corrosion rates (T4-26, T4-29; Table 10). The redox potential also can alter the speciation of dissolved chemicals and consequently the solution concentration of the corresponding elements. This effect has been prominent in the mild steel-lined Hanford waste storage tanks that contain a mixture of different phases having different chemical species. The rate of tank-wall corrosion is affected by the redox potential defined by presence and radiolysis of nitrate, nitrite, and hydroxide in the waste. Plutonium solubility is also affected by radiolysis of these same three species and its solubility can be sharply increased by chemical oxidants [42, 43].

As another example, the stress corrosion cracking (SCC) of sensitized Type 304 stainless steel (SS) in the primary coolant circuit of BWRs is entirely driven by the high redox potential of the coolant, which is caused by the accumulation of O and H_2O_2 through radiolysis and the loss of H through coolant boiling in the core. The redox potential and subsequent effects on SCC have been successfully determined in BWR primary coolants in which at least four redox couples are present [41]. This work is the basis for specifying a critical, primary circuit corrosion potential of -0.23 volts based on the standard H electrode (V_{SHE}). For potentials more negative than this value, sensitized 304 has been demonstrated to be resistant to SCC [41].

Several of the panelists expect radiolysis to begin quickly, during the ECCS injection phase (T2-20, Table 8). They stipulate that H_2O_2 production will begin almost immediately after the break, because the H control within the RCS will be lost, and much of the initial dissolved H in the RCS coolant will transition to gaseous H upon exposure to the containment atmosphere (T1-5, Table 7). As a result, Fe, Ni, and Co (and possibly other metals like Al) will be more easily oxidized and will precipitate oxides, oxyhydroxides, and hydroxides.

The presence of H_2O_2 in solution will promote further oxidation of $Fe(OH)_2$ to Fe_3O_4 or to iron oxyhydroxide (FeOOH). However, local depletion of H_2O_2 could also cause $Fe(OH)_2$ oxidation to Fe_3O_4 and H through the Schikorr reaction. The H thus generated would contribute to the total H inventory, which would help sustain reducing conditions as discussed in Section 5.1.4. The formation of perborates (BO_3^-) from the reaction of sodium metaborate ($NaBO_2$), NaOH, and H_2O_2, is also possible under radiolytic conditions [44]. Perborates are oxidizing agents and could therefore oxidize metals.

Because most participating redox reactions will be in a state of disequilibrium, the oxidation state of the system can only be defined in terms of mixed potentials.

The PIRT panelists expect radiolysis to be most significant during the early recirculation phase, from 20 minutes to approximately 1 day (T3-6, Table 9). Subsequent to this, radiolysis is expected to continue, but equilibrium may eventually be reached (T4-38, Table 10). As mentioned previously, high redox potentials can substantially elevate the metallic corrosion rates and alter chemical speciation and the solubility of compounds. Several panelists believe that the radiolytic effects on the redox potential could fundamentally impact the containment pool chemistry. However, other panelists believe that the oxidizing potential of the containment pool is already high which may mitigate the severity of this effect.

Another corrosion rate enhancing mechanism (for Al, Fe, and SS) promoted by radiolysis is the formation of hypochlorite or hypochlorous acid by radiolysis of Cl-bearing water (T4-29, Table 10; S-4, Table 12). Both of these compounds are strong oxidants. The PIRT panelists expect that effect could become active during the intermediate ECCS recirculation time, after 1 day or more. Most of the panelists believe that the impact of this effect is modest because containment pool Cl concentrations are expected to be low. The ICET test plan [12] indicated that a HCl concentration of 100 ppm (3×10^{-3} M), which corresponds to about 70 pounds of insulation degradation, represents Cl concentration released during cable radiolysis. Actual plant Cl concentrations could be lower than this value if significantly less cable insulation undergoes radiolysis. However, if Cl concentrations greater than 100 ppm (3×10^{-3} M) are present, hypochlorite formation could be significant. At least two panelists expect that the effect of radiolysis will be secondary for inorganic materials because the environment is already strongly oxidizing and that organic materials are more susceptible to radiolytic effects than inorganic materials (Section 5.6).

Although radiolysis induced by the activation product decay from the reactor fuel promotes global radiolytic effects, localized effects are possible because of radionuclide transportation and accumulation within the sump screen debris bed (T3-16, Table 9; T4-25, Table 10). This concentration of radionuclides (hundreds of Curies have been postulated) could locally alter the debris bed chemical environment. The radioactive flux becomes more concentrated than if it was dispersed throughout the containment pool, and the radiolytic reaction may not reach steady state as debris continues to accumulate. Two panelists believe that this concentration will create a locally higher oxidizing environment through production of H_2O_2. Reduced species will transform into oxidized species and increase the likelihood of precipitation or co-precipitation of oxides.

As with the global effects, two panelists stated that radionuclide concentration will minimally affect the inorganic chemistry and species formation, but it will have a greater impact on organic materials (Section 5.6). These panelists believe that local radiolysis will not significantly affect the already high oxidation state such that the corrosion rates and solid species precipitation observed in the ICET series would remain largely representative of the post-LOCA environment. However, radiation fields could cause organic materials to either decompose or polymerize and harden, depending on the material. Decomposed organics could also coat fiberglass insulation fibers to reduce leaching as the post-LOCA events continue. Another possible effect is that local radiation could cause glass embrittlement of NUKON® fibers, which would cause longer fibers to break.

Decay of the activation products may also contribute to the reducing potential of the post-LOCA solution through another mechanism. All radionuclide products formed in the reactor are negatron emitters. A negatron is a high-energy electron that originates in the nucleus. Although each individual radioactive decay yields only one electron, the interaction of the one highly energetic electron with water causes ionization events in water on the order of 1×10^4. In other words, one decay electron yields about 1×10^4 electrons in water. Gamma emitters also contribute to electron production. The presence of kiloradians worth of radioactive beta emitters in the containment pool will also produce significant numbers of electrons that may shift the redox potential towards a reducing environment—but only at interfaces between solid and aqueous phases—as compared to the same environment without decay products. However, this effect is expected to be beneficial because an environment that is less oxidizing will reduce metallic corrosion and leaching rates.

Traditionally, however, modeling of redox potential effects has considered only the chemical radiolysis products that are produced (e.g., H, H_2, OH, H_2O_2, HO_2, and O_2), because these are the longer lived species that form as a result of electron bombardment and capture. Also, only these species, and not electrons, can participate in charge transfer reactions at the surface that leave the two phases (i.e., aqueous and solid) electrically neutral. The number of particles of each radiolysis product in the system, produced upon the absorption of 100 eV of energy from the primary radiation (α, β, γ, and n radiation), is specified by the G-value.

Another related observation is that electron emission can occur from a negatively polarized metal surface in contact with an aqueous environment. However, the aquated electrons, so produced, appear to remain closely associated with the metal surface, probably for charge reasons, whereas the chemical radiolysis products are free to disperse throughout the aqueous phase. Thus, the overall effect of this electron emission is not expected to enhance or diminish the redox potential in the containment pool.

The previous discussion has focused on the radiolysis of water and the effects on the global and local redox conditions, but radiolysis also converts atmospheric nitrogen (N_2) into nitric acid (HNO_3) within the containment pool (S-1, Table 12). The HNO_3 acts to decrease the pool pH. One panelist stated that this effect, if the radioactive dose rate is sufficiently high in the containment pool, could overwhelm the plant's buffering capability. This panelist also conducted initial conservative scoping calculations that predicted a precipitous pH drop caused by this effect that would lead to massive corrosion of structural materials [27]. Plant licensees are required to account for radiolysis, including that of atmospheric N_2, in designing the buffering system of the pool in an alternative source term analysis [45]. A method for calculating pH to account for this and other radiolytic phenomena is also described in [46]. However, these calculations do not consider the effects on pH from all the various debris types existing in the post-LOCA containment pool. It is also not clear if the calculation accounts for the variability of boric acid concentration and acidity as a function of temperature and cation concentration. In light of these additional factors, some additional evaluation of the calculation methods described in [46] may be useful to validate that the containment pool pH does will not become acidic as a result of HNO_3 formed by the radiolysis of atmospheric N_2.

5.2.2 Effect of Radiation on the Properties of Solid Species

Another possible effect is that these radiation fields may alter the types and properties of solid phases that form (T4-25, Table 10). Such characteristics that could be affected by the radiation dose include the precipitate size, composition, morphology, and aging mechanisms. A specific example that may illustrate the effects of radiation on solid species is the observed decomposition of the neutron poisoning material (Boraflex®) in PWR spent fuel pools [47]. The borated, spent fuel pool pH is approximately 4.5. In the pools, reactive silica is created by the degradation of the silicone rubber polymer and silica filler material that comprises Boraflex®. However, similar Boraflex® degradation has not been observed in long-term testing (i.e., approximately 6 months) in either borated, aqueous environments or under radiation exposure in dry air environments, although it is not reported if the solution was analyzed for silica or other degradation products. The implication is that the degradation observed in the spent fuel pool results either from radiation exposure or from radiolysis-induced oxidants like H_2O_2 within the borated solution, and not from simple dissolution of the silica filler material.

The form of the dissolved silica may also be affected by aqueous, radiation exposure. It is hypothesized that the radiation from the spent fuel causes the silicone rubber polymer to hydrolyze and, together with the silica filler, dissolve as reactive silica. In spent fuel pools (pH approximately 4.5), reactive silica is present in solution at concentrations up to about 110 ppm. In contrast, the solubility of reactive silica in water (pH approximately 7) is about 51 ppm (1.8×10^{-3} M) as Si or 110 ppm (1.8×10^{-3} M) as silicon dioxide (SiO_2) [48]. Although solubility is affected by pH and other variables independent of radiation, it is also possible that the solubility has been affected by the radiation by changing the physical characteristics (e.g., precipitate size, composition, and morphology) of he silica. More study would be necessary to validate this hypothesis and verify the actual effects of radiation on the decomposition of Boraflex® in this example, and other debris in the post-LOCA containment pool. Some panelists also believe, based on similar arguments presented in Section 5.2.1, that the effect of radiation on the form and properties of organic materials could be more significant than any effects on inorganic material properties.

5.2.3 PIRT Research Recommendations

The ECCS performance implications associated with all the radiological issues discussed in this section are highly uncertain at this point. More consideration of radiological effects is warranted to first understand the magnitude of the radioactive fields that could be produced. Then, the effects of changes in the redox potential on the containment pool chemistry, material corrosion rates, and chemical speciation and solubility could be evaluated. Also, the likelihood of transport and accumulation of activated species to the sump screen debris bed should be assessed to determine the radiation field which could be concentrated at the sump screen. Any localized chemical effects that may result from radionuclides trapped within the debris bed and the effect of radiation on the properties of precipitated solid species could then be explored.

5.3 Physical, Chemical, and Biological Debris Sources

This section describes the types and quantities of debris that is created in the post-LOCA environment. Debris sources include physical debris (e.g., fibrous insulation, particulate insulation, and latent debris), chemically induced debris (e.g., oxides and precipitates), or debris created by biological growth. Debris generation is an important consideration in any post-LOCA analysis of ECCS operability because it defines the amount of debris that will exist in the post-LOCA containment environment. The ECCS must contend with the debris load that is transported to the sump screens and either accumulates on the screen or is ingested into the ECCS pumps. Conservative guidelines have been established to analyze and address many aspects of debris generation and most physical debris sources [4]. This section will focus on important effects related to debris generation and debris sources that are not currently considered in existing guidelines, or that have not been fully evaluated in NRC-sponsored or industry-sponsored research activities. These effects generally result from or are related to chemical or biological processes that may be active in the post-LOCA environment.

5.3.1 Debris Generation

The PIRT raised several issues related to physical debris generation, which may alter the post-LOCA physically induced (Section 5.3.2) and chemically induced (Section 5.3.3) debris. The first relates to the initial impingement of the two-phase jet emanating from the broken pipe (T1-6, Table 7). This, of course, is one important source for physical debris generation. Physical debris is also generated by containment spray actuation, and post-LOCA pressure/temperature transients within the containment building and within the reactor coolant loop.

Initial guidance for calculating post-LOCA debris generation is largely based on single liquid phase jet testing [4]. However, limited validation of insulation and coating debris damage or debris sizes from more representative two-phase jets [49] has suggested that the existing guidance is generally conservative. The size distribution of the insulation debris can affect both transport properties and chemical reactivity. Smaller sizes enhance transportability and reactivity [50 and 17]. In the absence of additional validation testing, the guidance requires that small debris sizes be assumed. Conservative debris concentrations are also calculated using the guidance unless corrosion inhibition or other advantageous effects (see below) can be demonstrated.

The debris caused by the jet resulting from metallic erosion, or the ablation of other materials like concrete, is not specifically considered in the guidance [4]. This phenomenon will govern the contributions of these materials in the early post-LOCA time period, before corrosion and leaching becomes important. Jet impingement could also initiate pitting corrosion, which could accelerate the corrosion of normally passivated materials like SS. However, the PIRT panelists generally recognized that corrosion, erosion, and ablation caused by jet impingement (e.g., Section 5.1.1) will likely be inconsequential because of the short time duration of the jet (i.e., approximately 30 s) and the relatively small affected volume relative to the rest of the containment.

The pipe break location is another important consideration. Breaks in different locations will create different debris characteristics such as the total debris mass, the particle/fiber mixture, and the debris composition. Of course, the physical debris characteristics affect the size (thickness), compressibility, and head loss associated with the physical debris bed that forms on the sump screen [51]. This phenomenon is the focus of the current guidance [4] that requires licensees to consider both high-fiber and high-particulate loading to demonstrate that acceptable NPSH margins are maintained in light of expected variability in the actual post-LOCA conditions.

Analogously, significantly different quantities of material (e.g., Cal-Sil and fiberglass insulation) can fundamentally alter the chemical effects (T1-9, Table 7). This impact was observed upon comparison of ICET Tests 1 and 4. The biggest difference between these tests was the inclusion of Cal-Sil insulation in ICET Test 4, yet substantial submerged Al corrosion occurred in ICET Test 1, whereas very little occurred in ICET Test 4. The lack of corrosion in ICET Test 4 was caused presumably by Al surface passivation by silica [14].

This result implies that, as with physical debris combinations, it is also necessary to consider the chemical ramifications arising from different break sizes and locations. A conservative or bounding analysis would not credit corrosion inhibition or other synergistic effects that reduce chemical effects. Such an analysis should need to consider only break scenarios that generate the maximum debris concentrations (e.g., insulation and latent debris) that contribute to post-LOCA chemical effect(s). These scenarios may or may not correspond to scenarios that generate the most physical debris or the highest ratio of particulate/fiber loading as required in the current guidance [4].

Conversely, if inhibition or other synergistic effects are to be credited for decreasing certain chemical effects, it is not sufficient to merely consider break scenarios that generate conservative concentrations of the debris that contribute to chemical effects. This practice may mask other, more realistic chemical effects. In a realistic analysis, break scenarios that generate less debris concentrations may induce the most onerous chemical effects

The following examples illustrate these points. A hypothetical plant using NUKON® fibrous and Cal-Sil particulate insulation could expect the greatest dissolution of Ca and Si by considering break scenarios resulting in high concentrations of each constituent within the containment pool. If this plant uses TSP buffering, then this scenario should produce conservative concentrations of $Ca_3(PO_4)_2$ precipitate. However, if the plant used NaOH buffering and has substantial submerged Al, this scenario may actually aid Al surface passivation and diminish the $Al(OH)_3$ chemical byproducts that form [11]. For a NaOH-buffered plant that wishes to credit Al surface passivation, it may be more appropriate to consider an alternative break scenario that considers lesser quantities of Cal-Sil particulate insulation. This scenario may lead to more significant Al corrosion and the formation of $Al(OH)_3$ caused by insufficient dissolved SiO_2 to induce Al surface passivation.

As another example, consider a hypothetical plant having Al reflective metal insulation (RMI)[2] that uses NaOH buffering. The most challenging chemical effects debris mixture could result from a break scenario producing sufficient fibrous debris to form a contiguous layer across the sump screen and the greatest quantity of Al RMI that settles within the containment pool. In this scenario, little additional particulate may be needed to cause head loss caused by the formation of $Al(OH)_3$ as demonstrated in the ANL testing [15].

These examples demonstrate that it is equally important to consider break scenario conditions, which result in the greatest challenges caused by chemical effects, separately from scenarios that may result in the most challenging fibrous/particulate

---

[2] Most reactor buildings in PWR plants do not contain Al RMI.

physical debris mixtures. Small variations in the containment pool chemical constituents are likely not significant assuming that contributions of all containment materials (e.g., eroded concrete, fiberglass, paint and polymer debris, and metal corrosion products) are appropriately considered. What is important is considering breaks that generate significantly different concentrations or combinations of materials.

The PIRT panelists also recognized the importance of containment spray actuation in transporting post-LOCA-generated insulation debris, latent debris, and coating debris to the containment pool (T2-7, Table 8). Submerged insulation debris within the containment pool, as demonstrated in the ICET tests [11], is expected to be a principal contributor of dissolved chemical species that contribute to various chemical phenomena. The latent debris concentration should be limited in those plants that have improved cleanliness procedures. Such procedures are identified as a possible interim compensatory measure [52] to reduce the potential risk caused by post-accident debris blockage as the generic letter evaluations [3] are proceeding. Therefore, latent debris may only be significant for plants that have little other fibrous or particulate debris sources. Contributions to debris from coating are expected to vary tremendously among plants as a function of the relative amounts of qualified and unqualified coatings in containment. The initial jet impingement emanating from the pipe break is assumed to fail both qualified and unqualified coatings within each respective zones of influence (ZOI) [4]. However, qualified coatings are assumed to remain intact outside of their ZOI, whereas unqualified coatings are assumed to fail outside of their ZOI [4]. Therefore, a substantial quantity of unqualified coating debris could be produced.

The current guidance for physical debris transport is conservative in that bounding quantities of physical debris are assumed to be transported to the containment pool [4]. However, as with the rationale for considering multiple break scenarios, the maximum quantity of physical debris at the sump screen may not result in the most deleterious chemical effects. Therefore, it is important in a realistic chemical analysis to consider scenarios where physical debris may be sequestered from the containment pool as a function of the break location, debris location, and the containment configuration (T1-9, Table 7). Debris that may significantly contribute to either advantageous or deleterious chemical effects (e.g., Cal-Sil) may be more amenable to sequestration. For example, silicates could lead to inhibition of Al corrosion if present in sufficient quantities [53]. However, inhibition might not occur if substantial Cal-Sil or other silicate debris sources did not reach the containment pool. As with debris generation considerations, conservative analyses that do not credit either debris sequestration, corrosion inhibition, or other advantageous synergistic chemical effects may only need to evaluate scenarios that generate the maximum debris types and quantities that may contribute to chemical effects.

5.3.2 Physical Debris

As mentioned in Section 5.2, activated species that spall from reactor component internals and fuel represent a particulate source. One principal activated species source is from Fe and Ni oxides (mixtures of nickel ferrite, magnetite, nickel oxide and hematite), more commonly called "crud," formed during corrosion of the RCS piping The crud layer is approximately 125 μm thick. It is postulated that these oxides could be dislodged by the hydraulic shock associated with the pipe break (T1-1, Table 7). These materials are observed during normal shutdown transients that produce less severe hydraulic shock. Plant data provided by one of the PIRT panelists for three reactor shutdowns indicated that Ni concentrations of approximately 9 ppm (2×10^{-4} M) were observed between 30 and 50 hours after shutdown. The Fe concentrations were approximately 0.6 ppm (1×10^{-5} M) and occurred between 15 to 20 hours after shutdown [47]. These concentrations were corroborated by a study of French PWR shutdown chemistry that also documented the release of ^{58}Co and ^{60}Co [54].

One panelist postulated that up to 100 ppm (approximately 2×10^{-3} M) of Ni and Fe could be released just after the pipe break as a result of the more severe LOCA pressure and temperature shock. The following postulated scenario (T2-13, Table 8); that could lead to these concentrations The initial oxide layer breaks off, fresh metal is exposed and the exposed reduced Ni and Fe combines with air and forms additional oxides that spall off because of the continued pressure/temperature fluctuations. Thus, the cycle is repeated. This process could continue for 1 day or longer until RCS piping cools down. This process may also be enhanced if either hot STB or boric acid acts as a flux to dissolve the Fe_3O_4 layer from the reactor vessel wall or zirconium oxide (ZrO_2) from the clad fuel. However, the RCS temperature may be too low to initiate this phenomenon [27]. If these crud particulates are transported to the sump screen, they could contribute to NPSH reduction from additional particle loading. Also, these Fe and Ni activation products could concentrate at the sump screen and may result in local radiation dose rates on the order of kiloradians per hour. This radiation flux could then alter the local chemical environment and solid species formation and properties (Section 5.2). Finally, over the course of the post-LOCA mission time, the crud materials may change their chemical form, including the formation of ferric hydroxide (a strongly flocculating material) that could produce deleterious chemical products.

However, at least one panelist believes that although the initial crud burst quantities may be appreciable, subsequent concentrations caused by continued generation and spallation will be relatively small compared to the other solid species present within the containment pool. A few panelists also believe that the crud particles will settle in a relatively quiescent pool given their high densities. This would diminish radiation concentration and particulate loading effects at the sump screen. Another panelist suggested that a positive effect of these crud particulates is that their high density will promote agglomeration of other insoluble materials.

Spallation of the reactor core fuel cladding ZrO_2 (T4-24, Table 10) or deposited chemical products is another potential source for activated material (T3-22 and T3-27; Table 9; T4-37 and T4-38; Table 10). Possible post-LOCA chemical products that could be deposited on the fuel clad include organics, Al, B, Ni, Fe, Zn, Ca, magnesium (Mg), silicates (SiO_3^{2-} and SiO_4^{4-}), and CO_3^{2-}-based deposits. Solid species could precipitate on the fuel cladding because of retrograde solubility (Section 5.4.2.2) or solvent evaporation near the fuel. Additional solids that form either in the containment pool or heat exchanger (Section 5.4.2.1) will also be present. Note that most of the preexisting solids, especially those that would be transported through the sump screen, are expected to be fine particulates. These products could bind on the fuel cladding as the coolant water bearing the dissolved and suspended solids flashes to steam. The deposited solids could then undergo higher temperature hydrothermal reactions and self-cementation or caking. The characterization of crud found on PWR fuel cladding from operating reactors [55] provides evidence to support this hypothesized post-LOCA fuel deposition scenario. The referred study found organic and aluminosilicate deposits in addition to nickel ferrites on fuel cladding surfaces.

As the deposits thicken with time, thermal shock and mechanical vibration caused by the post-LOCA recirculating water could cause the deposits to spall (T4-38, Table 10). As the fuel surface continues to cool, spallation from the surfaces will continue. This phenomenon is observed in spent reactor fuel pools as samples from the bottom of the pools exhibit increased concentrations of spalled fuel deposits over the first several months of storage [47]. Spalled fuel deposits could either contribute to clogging within the reactor core (Section 5.7.3.2), or they could be transported to and contribute to sump screen head loss. Transportation may be aided if the spalled deposits are particulates or are curved flakes resulting from the fuel shape. A positive consequence of spallation is that it will improve the heat transfer of the reactor fuel (Section 5.7.3.1). However, at least one panelist believes that spalled deposits from the reactor fuel are not likely to generate a significant amount of solid material compared to the other chemicallyinduced debris constituents.

5.3.3 Post-LOCA Chemically Induced Debris

5.3.3.1 CO_2 Absorption

Much like physical debris that is generated at the onset of the LOCA, chemically induced debris is expected to form early, and throughout, the post-LOCA scenario. Air entrainment within the containment pool—caused by coolant water emanating from the break and pool turbulence in addition to atmospheric scrubbing from containment spray (T2-8,Table 8) —will cause carbon dioxide (CO_2) absorption beginning soon after the LOCA and before the onset of sump recirculation. It is worth noting that the large containment-air-to-pool-volume ratio that exists in plants was not simulated in the ICET program. With the dissolution of post-LOCA debris (e.g., Cal-Sil, concrete, and fiberglass), Ca cations (Ca^{2+}) would build up in solution. Because the pH-buffered recirculating cooling water eventually becomes alkaline (i.e., the pH is between 7–10), the absorbed CO_2 would accumulate predominantly as the bicarbonate ion, which would, in turn, react with Ca^{2+} to precipitate $CaCO_3$. Both the absorption of CO_2 and precipitation of $CaCO_3$ would tend to depress pH, which, together with the depletion of Ca^{2+} from solution, would enhance the continuing dissolution of insulation materials. However, the pH buffers in solution would mitigate pH changes, and eventual depletion of CO_2 in the air of the containment building would limit further calcite precipitation.

Preliminary calculations by some of the PIRT panelists suggest that approximately several hundred pounds of finely divided $CaCO_3$ particulate may be produced through the uptake of atmospheric CO_2 in a containment building (T2-10, Table 8). This quantity is likely negligible (i.e., less than 1 percent) of the total debris introduced into the containment pool following a LOCA. Additionally, depending on the buffering chemical, the $CaCO_3$ may simply replace other finally divided Ca precipitates like $Ca_3(PO_4)_2$. The hydrated lime, $Ca_3(PO_4)_2$, and calcite solids are physically similar and are unlikely to be significantly different in affecting either sump screen flow or downstream ECCS performance. Therefore, calcite formation may only be a principal consideration if conditions do not foster precipitation of these other species and if other debris loading sources are small (i.e., less than a few thousand pounds).

However, the $CaCO_3$ could assume a secondary role by serving as a site for heterogeneous nucleation and growth of other chemical phases (e.g., Ca phosphates and silicates) early in the post-LOCA cooling cycle (T2-10, Table 8). Thus, although calcite from atmospheric CO_2 may produce a negligible amount of solid material, its effect on precipitation kinetics could be significant. For this reason, a few PIRT reviewers recommended that the relative impact of CO_2 absorbed from the air and the effect of subsequent calcite formation should be assessed. The accurate assessment of these implications requires consideration of plant-specific debris sources, dissolved Ca levels, sump environmental conditions, and the quantity of fresh air in containment that replenishes the CO_2 supply that may have been exhausted by precipitation.

5.3.3.2 Dissolution and Corrosion

Debris dissolution and metallic corrosion also would begin just after the LOCA as materials are impacted by and then immersed within the containment pool water. These are the major mechanisms for creating the ionic species existing in the containment pool (T2-9, Table 8). The dissolved, ionic species could subsequently react and precipitate to form new, solid phases that were not originally in the containment pool. Therefore, corrosion and dissolution will determine the total inventory of solid material in the containment pool. The importance of metallic corrosion and debris dissolution is well known and has been demonstrated in the ICET program for representative post-LOCA environments [11] and also in less-complex, separate effects testing [17 and 18].

The focus of this prior research has often been the evaluation of conservative amounts of representative containment materials. It may also be necessary to study the corrosion and/or dissolution and precipitation for realistic amounts of important plant-specific materials in simulated post-LOCA environments. The chemical products determined from conservative material concentrations may suffice for assessing a dominant chemical reaction that is not expected to be altered significantly by other chemical additions. In other situations, using conservative material concentrations may mask more likely chemical effects that could alter the solid species and phases that form.

For instance, ICET Test 3 included a bounding amount of Cal-Sil insulation that led to the rapid elimination of dissolved PO_4^{3-} ions (within 1 day) through the formation of $Ca_3(PO_4)_2$. Although it was demonstrated that $Ca_3(PO_4)_2$ can result in significant head loss [15], it was not possible in ICET Test 3 to form other PO_4^{3-} products that may have different implications for environments containing much less dissolved Ca from Cal-Sil or other sources.

Knowledge of realistic chemical mechanisms also allows more accurate evaluation of the time period associated with possible solid-species precipitation. The minimum NPSH pump margin and maximum required ECCS cooling rates occur at the onset of recirculation and represent the principal design considerations. Greater system margin is available as time increases past the onset of recirculation. The ICET Test 3 [11] and subsequent head-loss testing [15] demonstrated that significant quantities of $Ca_3(PO_4)_2$ can form within an hour of the initiating LOCA event. However, in ICET Tests 1 and 5, significant quantities of $Al(OH)_3$ solid was not observed until several days had passed. Therefore, accurate consideration of precipitation formation could be used to demonstrate that an acceptable margin exists. Conservative assumptions of the time to form precipitates (e.g., instantaneously) could always be used in the evaluation as long as the actual solid species precipitates, as described in the above paragraph, are known.

5.3.3.2.1 Nonmetallic Dissolution

The implications on ECCS performance associated with dissolution of nonmetallic physical debris within the containment pool are not always straightforward. On one hand, leaching may reduce the debris mass that contributes to sump screen head loss if the leachant remains dissolved. This phenomenon was observed in mixed fiberglass, Cal-Sil debris beds in STB environments [15]. Calcium silicate dissolution decreased the measured head loss across the test screen after some time. Conversely, some fiberglass or particulate dissolution may allow greater physical debris bed compression which can increase head loss (T3-18, Table 9). Leachants (e.g., Si, Al, Mg, Ca, and cations or oxyanions) may also react with other components in the containment pool to form voluminous gel-like products that may have onerous head loss and downstream effect implications (T3-7, Table 9). This issue underscores the need to consider the synergistic effects created by debris dissolution and metallic corrosion.

The effect of aging on the leaching process for nonmetallic materials, especially concrete, is another consideration (S-3, Table 12). Ordinary Portland cement reacts with water to form principally calcium silicate hydrate (C-S-H) (i.e., $CaO-SiO_2-H_2O$) "gel" and portlandite ($Ca(OH)_2$). It is these phases that should be considered if fresh concrete were to be leached by the recirculating coolant. However, neither the exposed concrete faces nor concrete dust in the containment building are likely to be fresh. After 30 years of atmospheric exposure, a substantial fraction, if not all, of both the exposed C-S-H gel and the portlandite will be carbonated [56]. Exposed carbonated cementitious materials would most likely be converted to a mixture of amorphous silica and either vaterite or aragonite, both forms of $CaCO_3$ [57].

5.3.3.2.2 Metallic Corrosion

The implications on ECCS performance associated with metallic corrosion are similarly complex. Bulk Al corrosion and the formation of $Al(OH)_3$ precipitates upon cooling were observed in ICET Test 1 [11]. These precipitates were later demonstrated to have significant head loss implications [15] at the ICET concentrations. The ICET Test 5 also exhibited significant Al corrosion. However, significant Al corrosion was not apparent in ICET Tests 2, 3, and 4. Additionally, significant corrosion of other metallic components was not observed in any of the ICET tests. See Section 1.2 for more details.

Although all panelists agree that Al corrosion is important in the post-LOCA environment, a few believe that higher Al corrosion rates are possible if affected by galvanic coupling, by electrolytic deposition, or by catalysis of other metals like copper (Cu) (T4-6, Table 10). Copper was included in the ICET program, which may imply that the electrolytic deposition effect, as described below, or any potential catalytic effect has already been accounted for at the Cu surface area to water volume ratios used in ICET. Higher Cu ratios may therefore be required to trigger these effects. The catalytic effect of greater amounts of Cu, or other materials not in the ICET program, has not been studied.

The panelists do not expect Cu by itself to corrode significantly unless the post-LOCA environment is highly oxidizing (T4-3, Table 10). This expectation was confirmed in the ICET series where minimal Cu corrosion was observed. However, besides catalytic effects mentioned above, Cu can accelerate or inhibit corrosion of other metals.

One way in which Cu can alter the corrosion rates of other materials is by forming a galvanic couple (T2-22, Table 8). A galvanic couple is created when a structural material and Cu are in electrical contact and submerged within an ionic solution. Less noble metals than Cu can then preferentially corrode through anodic dissolution. For the prominent containment materials, the materials ranked by decreasing nobility are as follows: Cu, Fe, Zn, and Al. Thus, Al is more likely to corrode than other materials and can actually inhibit corrosion in other materials if a galvanic coupling is present. Copper is used, uncoated, as the grounding straps for many large structural components. Therefore, the possibility for a galvanic couple between these components exists if they are submerged during a LOCA. One panelist believes, however, that precipitates could be attracted to the Cu grounding straps and form a coating that would diminish the galvanic couple. Galvanic effects have not been studied in the ICET series or in other related research projects (Section 1.2).

A mechanism that decreases corrosion rates is Cu deposition on the surface of other materials. Copper ion deposition was observed on Al coupons in ICET Tests 2 and 3, and the measured Al solution concentration was below the detection limits in these tests [11]. Therefore, corrosion inhibition through Cu ion deposition may have occurred. It is difficult to know which, if any, of these possibly competing Cu effects are important without further analysis.

Iron, or steel, corrosion was not particularly prevalent during the ICET series and other testing in alkaline environments (T4-4, Table 10). Under the oxidizing conditions present in the ICET series, the dissolution of metallic Fe (i.e., in steel) might be expected to lead to immediate oxidation with precipitation of Fe oxides under oxidizing alkaline conditions as oxidation of Fe^{2+} is catalyzed by OH^- [58 and 59]. However, Fe passivation could occur if Fe_3O_4 precipitates on exposed steel surfaces under alkaline conditions, as is suggested by [13]. This mechanism could explain why little Fe corrosion was evident during the ICET series. At lower pH (i.e., pH of approximately 7), Fe^{2+} is expected to accumulate in the aqueous phase, but the presence of O during ECCS coolant recirculation should ensure that any Fe precipitate would occur in the Fe(III) state. The PIRT panelists do not expect that Fe would precipitate in the Fe(II) state as reported in [60].

Lead (Pb) is an element that was not included in the ICET series, but it was evaluated by the PIRT (T4-2, Table 10). Lead blanketing or Pb wool is used to shield radiation hot spots during containment outage operations. One panelist estimated that several hundred pounds of Pb may be used in blankets. Although the blankets are usually covered with a protective plastic coating, it is expected that the coating will be penetrated in many places as a result of normal operational wear and tear. However, much of this Pb should be removed after an outage. Additionally, under typical post-LOCA containment conditions, Pb corrosion is expected to be low (on the order of ppb to low ppm). Passivation of the surface through formation of complex Pb silicates is also possible.

Acetate ($CH_3CO_2^-$) present within containment could enhance Pb dissolution and cause lead carbonate percipitation within the containment pool. The PIRT panelists largely agreed that the Pb chemical contributions were extremely plant-specific and these effects are only potentially significant if large quantities of both unprotected Pb (greater than several hundred pounds) and $CH_3CO_2^-$ are present in the containment pool. Lead can also induce cracking in SS components, but this mechanism should not degrade ECCS operability over 30 days.

Another variable not simulated within the ICET test series is the behavior of silica in the RWST and the RCS (S-2, Table 12). The concentration of silica in the RWST may be as high as 3 ppm (5×10^{-5} M), and in the RCS the silica concentration will be in the range of

0.3 to 3.0 ppm (0.5 to 5×10^{-5} M). Because silica in combination with Mg, Ca and Al can form materials with retrograde solubility, silica can have an effect on the total mass of material precipitating. The dissolved silica could react with other species leading to the formation of secondary precipitates. At least one PIRT panelist believed that these contributions could be important. Although this concentration of silica in the RWST should be considered when accounting for dissolved silica levels, it should be noted that silica levels in the ICET series generally exceeded 10 ppm (2×10^{-4} M) for all tests. Therefore, the additional contribution resulting from RCS and RWST silica may be insignificant compared to silica from other containment debris sources.

5.3.3.2.3 Dissolution Rate Inhibition and Acceleration

Other than the ICET program, most of the characterization and analysis to date in the post-LOCA environments has used separate effect testing to evaluate corrosion. The complex interaction between dissolution processes of combined, or multiple, materials or second order effects resulting from variations in the trace chemicals of the containment water have not been addressed. Examples of synergistic dissolution rate effects observed in the ICET series include the inhibition of NUKON® fiber leaching by dissolved Al, inhibition of Al corrosion by dissolved silica (T4-8, Table 10), and possible suppression of Al corrosion by Cu deposition. Another possible synergistic effect discussed previously is galvanic corrosion (Section 5.3.3.2.2) and concrete aging (Section 5.3.3.2.1). Some other examples of synergistic dissolution rate effects that the PIRT panelists believe may be relevant (e.g., S-4, Table 12) within the post-LOCA environment follow.

The impact on the Al corrosion rate caused by typical water constituents has been previously studied [61]. The constituents whose effects were evaluated included Cl, nitrate (NO_3^-), sulfate (SO_4^{2-}), carbonate/bicarbonate (CO_3^{2-}/HCO_3^-), H_2O_2, $CH_3CO_2^-$, oxalate ($C_2O_4^{2-}$), citrate ($C_3H_4OH(COO)_3^{3-}$), arsenate ($HAsO_4^{2-}/H_2AsO_4^{1-}$), phosphate ($HPO_4^{2-}/H_2PO_4^{1-}$), silicates ($SiO_4^{4-}$ and SiO_3^{2-}), chromate/dichromate ($CrO_4^{2-}/Cr_2O_7^{2-}$), molybdate ($Mo_7O_{24}^{6-}$), and mixtures of these species. Individual tests that were conducted at 92 °C (198 °F) with 100 ppm of the respective constituents, showed that only citrate increased corrosion rate, whereas phosphate decreased the corrosion rate. Significantly, H_2O_2 had no effect on the Al corrosion rate. Additional testing of Cl concentrations up to 10,000 ppm (0.3 M) and mixtures of 100 ppm Cl⁻ (3×10^{-3} M) and 100 ppm H_2O_2 (3×10^{-3} M) also had no effect on the Al corrosion rate.

Further Al corrosion tests have studied corrosion inhibition from phosphate [61 and 62] (S-4, Table 12). Initial work showed that measurable corrosion inhibition occurred with as little as 5 ppm (5×10^{-5} M) phosphate and postulated that the inhibition occurred because of surface deposition of a protective layer of aluminum phosphate [61]. Subsequent X-ray diffraction studies of an Al surface exposed to 5 ppm (5×10^{-5} M) phosphate at 195°C (383°F) for 24 days confirmed the formation of the aluminum phosphate compound augelite, $Al_2PO_4(OH)_3$. This surface compound forms at phosphate concentrations down to 0.5 ppm (5×10^{-6} M) [62]. Some salts including chromates, dichromates, silicates (as in ICET Test 4), and borates are also reported to inhibit Al corrosion [63].

Material alloying may also influence corrosion rates (S-3, Table 12). Corrosion data for various containment metals was extracted from the published literature and tabulated [60]. The corrosion rates varied widely and were a function of the specific corrosion conditions and the nature of the alloy subject to corrosion. Numerous different Al (and other metallic) alloys are commercially available and in use, but the variability in alloy dependent corrosion rates and cation concentrations has not been addressed in the NRC-sponsored testing. Given the corrosive environment of alkali borate solutions, at least one PIRT panelist expects that the nature of the alloy may not significantly alter corrosion, at least for Al. This intuition is buttressed by industry testing of several common nuclear Al alloys [53], which demonstrated only nominal corrosion rate differences. However, this has not been demonstrated for other metallic materials. A related consideration is the cation concentration or other chemical effects resulting from the alloying constituents. This consideration may only be important if substantially different alloying constituents other than those tested in ICET and other related programs are present or if significant alloying additions are present.

These various studies indicate that the chemical constituents within the containment pool environment and material alloying can alter the basic corrosion potential of materials, either advantageously or disadvantageously. It will therefore be necessary to consider plant-specific differences in the baseline containment pool environment and containment materials to determine if corrosion accelerators or inhibitors may be present. An evaluation of significant alloying additions in nuclear materials and additional study (i.e., consideration of prior research or bench-top testing) could be performed to determine if either significant corrosion rate differences or dissolved concentrations of alloying elements are possible.

5.3.4 Biological Debris

The propensity for bacteria or other biota to grow in preexisting debris beds located on the sump strainer screen or elsewhere within the ECCS was considered by the PIRT (T4-22, Table 10). Significant bacterial growth may be important if it creates additional debris that contributes to sump screen clogging or detrimental performance of downstream components like pumps and valves. The scenario postulated by the PIRT is that containment environments contain a large amount of biological materials (i.e., fungi, bacteria, and spores) because refueling occurs typically in the spring and fall with the refueling hatch open and the containment building under negative pressure to prevent gaseous or particulate radioactive releases. Under normal operating conditions, these materials do not affect the performance of the closed nuclear systems.

In the post-LOCA period however, initial temperatures will not be sufficiently high to sterilize the interior of the reactor containment building. As a result, as the reactor and surrounding component temperatures decrease, the warm moist conditions, augmented in some cases by phosphate buffering, will be conducive to growth of algae and perhaps other life forms. Sump screen debris beds and other similar locations are preferential locations for trapping and concentrating the biological materials and the organic food sources necessary to sustain their life. The proximity of the materials and food sources may foster growth of products like polysaccharides that are sticky and tenacious. According to one panelist, polysaccharides are known to cause fouling in other power plant systems, especially on SS surfaces. With time, these biological materials may develop contiguous mats and participate in blocking flow through the sump screens.

The panelists did not reach a consensus concerning the plausibility and impact of this proposed scenario within a 30-day time period. One panelist questioned whether biological materials could even survive and grow in the post-LOCA borated (approximately 2500 ppm (0.23 M)) water solution. He noted that borates are used as algae stats to inhibit growth. There was also disagreement about the time and temperatures needed for significant growth if biological materials do form. One panelist believes that biological growth will begin immediately and could increase exponentially such that significant challenges to both heat transfer and flow may exist within 15 days. Other panelists believed that growth would occur much more slowly and not be important over a 30-day time period.

However, it should be noted that significant biological debris was found in the Three Mile Island (TMI-2) post-LOCA containment when it was opened which contradicts the above notion that biological materials may not survive or grow in these environments. However, river water was mixed with the RCS and ECCS water, which would not be expected in most post-LOCA PWR containment pools [9]. Also, this observation was made from containment water samples taken several months after the LOCA occurred [8]. It is unknown how much of this debris was present over the 30-day time period most critical for ECCS recirculation performance. Given the PIRT uncertainty and lack of direct observational experience, this is an area where additional study could shed light on the extent of biological growth within the post-LOCA environment.

5.3.5 PIRT Research Recommendations

A number of issues identified within this section should be considered to thoroughly evaluate both generic and plant-specific chemical effects related to debris generation and to physical, chemical, and biological debris sources. Several of these issues can be addressed by understanding existing chemical effects over the range of appropriate, actual post-LOCA conditions at the plant. This study can be used to identify which issues remain relevant. Other issues may require additional study to more fully understand the particular chemical effects in order to make a determination of generic or plant-specific relevance.

Although the debris generation guidance provided in [4] requires an evaluation of conservative concentrations of the main types of expected physical debris, one of the PIRT panelists believes that integrated testing should be performed to quantitatively measure effects associated with physical debris generation. Specially, the test should simulate the initial RCS two-phase jet to consider the debris generated during the initial 10 s of the LOCA from different insulation materials and components. Then debris sizes, degradation, and chemical effects should be characterized.

This type of integrated testing certainly represents one approach that could be attempted to more realistically simulate the post-LOCA environment and conditions. However, there are a large number of variables (e.g., debris type, distance from break, break size, plant configuration, environmental conditions, and debris age) that affect debris quantities, sizes, and post-LOCA effects. Additionally, there are large uncertainties associated with these variables and large plant-specific differences among these variables. These factors conspire to increase the complexity associated with this type of testing. Therefore, a conservative consideration of the applicability of existing debris generation and degradation (Section 1.2) may be more appropriate.

However, no prior testing has evaluated the physical debris concentrations and characteristics from crud production and reactor fuel clad spallation. Additional evaluations could be performed to examine the possibility and extent of corrosion product spallation from reactor vessel walls and from fuel cladding surfaces under conditions that would occur under post-LOCA thermal and mechanical shock. Spallation of both initial deposits and deposits that continue to form (Section 5.3.2) in the post-LOCA period should be considered when evaluating the total debris concentration. This activated debris concentration will also determine the radiation fields (Section 5.2) present within the containment pool and, more importantly, at the sump screen. Evidence of these deposits is apparent in post-shutdown analysis of reactor water chemistry and in spent fuel pools upon continued fuel cooling, but larger quantities could be expected in a LOCA because of the higher pressure/temperature shock.

Although the composition of reactor vessel crud deposits are well known (Fe and Ni oxides), there are a host of potential chemical precipitates that could form either in the coolant by retrograde solubility or on the reactor fuel clad surface by water evaporation. The nature of these deposits can first be determined through consideration of the dissolved species and suspended solids that are present within the ECCS cooling water. A thorough study will likely require a plant-specific evaluation of materials and various break conditions to characterize the ECCS cooling water chemistry. Then, the effect of the higher reactor core temperatures on the ECCS cooling water chemistry can be investigated to determine if new solid species form and whether solid species suspended within the cooling water can adhere and/or subsequently spall from the fuel.

Research could also be conducted to determine whether these products are likely to settle within the reactor core, the lower plenum of the reactor vessel, or the containment pool. Settling within the reactor core may result in additional flow blockage and impede fuel cooling, but settling is much more likely within the containment pool because of the generally lower flow velocities. Containment pool settling will reduce the debris load that could contribute to either sump screen clogging or subsequent downstream effects. Other implications from, and the possible study of, spalled reactor fuel clad deposits are discussed in Section 5.7.3.

Research is also needed to evaluate the debris generated by post-LOCA chemical effects. As discussed, several types of precipitates were observed in the ICET testing [11] that have been subsequently studied to evaluate some important properties (e.g., settling, solubility, and particle size) [15] and the potential for sump screen head loss [15]. However, other chemical effects were not considered or properly accounted for in the ICET.

Calcite formation from atmospheric CO_2 was one phenomenon not properly simulated in ICET because the free headspace to water surface area and volume were not scaled to containment ratios. Calcite precipitate loading may only be important for plants with significant sources of dissolved Ca (e.g., Cal-Sil, exposed concrete, and fiberglass insulation) that use a buffer other than TSP. Plants with TSP buffering already contend with $Ca_3(PO_4)_2$. However, for all plants, the calcite may also serve as a site for heterogeneous nucleation and growth of other chemical phases (e.g., Ca phosphates and silicates) early in the post-LOCA cooling cycle. Thus, although calcite from atmospheric CO_2 may produce a negligible amount of material in plants, its effect on precipitation kinetics could be significant. It may be necessary to consider plant-specific debris sources, dissolved Ca levels, and sump environmental conditions to evaluate the significance of this phenomenon.

The NRC-sponsored research also did not rigorously study the synergistic effects on metallic corrosion and nonmetallic dissolution under post-LOCA conditions. Single-effects testing under the premise of developing conservative corrosion rates has been instructive [18]. However, synergistic effects can either increase corrosion rates through catalytic effects or decrease corrosion rates through passivation or other mechanisms compared to these single-effects values. Integrated tests using generally conservative material quantities, such as ICET, provide some evidence of synergistic effects, but conservative concentrations of materials can also mask or inhibit corrosion of other materials in these tests. This phenomenon could alter the types and amounts of any precipitates that form in the tests compared to the actual containment environment.

Realistic, plant-specific containment pool environment, materials, and concentrations should be considered to identify the appropriate corrosion acceleration and inhibition factors. One acceleration factor that should be studied is the catalytic effect of other materials that are not in the ICET program but may exist within containment. Copper is one possible catalytic material, but it would need to be present in greater quantities than in the ICET tests because no catalytic effect was observed in these tests. Examination of Al corrosion acceleration caused by citrate and other hydroxyl-organic acid anions (e.g., glycolate and $HOCH_2CO_2^-$) is only a concern if significant sources of these anions are expected. Possible corrosion inhibition effects that could be evaluated include the aging of concrete and other nonmetallic materials; the presence of phosphate and salts such as chromates, dichromates, silicates, and borates on Al inhibition; and the possibility and threshold for corrosion inhibition or acceleration through Cu ion deposition. Although corrosion inhibition is largely beneficial, it will be important to demonstrate that testing or evaluation conditions are representative of service conditions so that the results are relevant and proper credit can be determined.

Another corrosion inhibition or acceleration aspect that should be considered is the effect of alloying additions in nuclear structural materials and debris within the containment pool. First, knowledge of the particular material form and composition is needed in the assessment. Then, consideration of prior research or additional bench-top testing can be used to determine whether significant corrosion rate differences occur compared to unalloyed materials evaluated in ICET [11] and single-effects testing [18]. This testing could augment the information developed for Al alloys in [53] and should also determine whether significant dissolved concentrations of alloying elements are generated through corrosion. These alloying additions could form, or aid in the formation of, solid species.

Galvanic effects should be also considered. These were not considered in the ICET series. Galvanic effects can accelerate corrosion of less noble materials while inhibiting corrosion of more noble materials. Galvanic couples existing within the containment pool (caused by direct contact between dissimilar metals) should be identified on a plant-specific basis. There is a wealth of existing information on galvanic effects [64], but limited testing of the strongest galvanic couples under simulated containment pool conditions could be used to verify the applicability of identified corrosion rates.

Finally, research would be useful in determining if biological growth is a viable post-LOCA debris source that has significant implications. Screening tests could initially be performed to determine if growth is possible in post-LOCA borated water solutions containing organic sources at representative temperature profiles. A wide range of borate concentrations, organic substances, and temperature profiles could be evaluated to most effectively bracket plant-specific conditions. If growth is demonstrated within 30 days, representative changes in important variables (e.g., flow rate) that may accelerate or retard the biological growth rate could be evaluated to assess possible quantities of biological materials that could form at the sump screen debris bed or other downstream locations.

5.4 Solid Species Precipitation

This section discusses several issues evaluated by the PIRT, which may affect the formation or precipitation of solid species from previously dissolved ionic species contained within the post-LOCA containment pool. Simply, precipitation occurs when the solubility of a specific solid phase is exceeded in solution. However, precipitation can depend on a complex interaction of multiple synergistic variables such as time, temperature, pH, and other species in solution. It should be noted that precipitation, as defined in this section, is the creation of small (nanometer to sub-micrometer size) particles suspended within the ECCS coolant. Issues related to the growth or agglomeration of these initially small particles are discussed in Section 5.5. Precipitation, along with growth or agglomeration, is likely needed to form solid species that are large enough to have significant implications for ECCS performance.

5.4.1 Post-LOCA Precipitation Mechanisms

Inorganic polymerization (T2-18, Table 8) is the process by which hydrolyzed cations link together, generally through oxo (-O-) bridges and dehydration, to form chains and networks. With sufficient growth, colloidal particles will result. Condensation polymerization occurs when the covalent bonds are rearranged such that monomers (the basic building blocks) are connected and water is expelled. The PIRT panelists expect the precipitation process to be a principal mechanism for the formation of solids in the post-LOCA containment pool, as demonstrated in the ICET testing [14]. This process may also be necessary to form particles large enough to tangibly affect ECCS performance. The PIRT panelists hypothesized the following possible scenario describing this process.

Polymerization may begin early in the post-LOCA scenario (within 20 minutes of the initiating event) as cations are introduced in the containment pool. Any of the cations that can induce precipitation in the post-LOCA environment are candidates for polymerization including Al, Fe, Si, Zn, Cu, Ca, and boric acid. Extensive information is available for the hydrolysis and precipitation of solids from solutions containing simple cations, such as Al^{3+} and Fe^{3+}, but less information is available for more complex systems (e.g., precipitation of aluminosilicates). The conditions of polymerization for these materials are a function of the ionic species and concentrations, time, temperature, and pH. It is therefore necessary to either realistically or conservatively consider applicable combinations of these principal variables to determine whether solid species precipitation and ripening through polymerization occurs. This evaluation could be useful in determining the type and concentrations of precipitates that form under plant-specific conditions.

Precipitation is fostered by the existence of heterogeneous nucleation sites within the containment pool (T3-15, Table 9). Conversely, a dearth of these sites can delay precipitation. Nucleation sites include dirt particles, coating and insulation debris, and biological debris. All panelists believe that it is a foregone conclusion that the containment environment would contain a plethora of heterogeneous nucleation sites to foster precipitation. Additionally, it is expected that standard laboratory environments and simulations should also contain sufficiently representative types and amounts of nucleation sites unless extreme cleanliness is practiced. Therefore, although these sites are important, they exist naturally and in both laboratory and post-LOCA environments.

As mentioned previously, classical precipitation will not occur unless the solubility limit for a particular solid species is exceeded. However, the co-precipitation of other species is possible below the solubility limit of those species (T2-19, Table 8). Co-precipitation occurs when a normally soluble ion becomes either included or occluded into the crystalline structure of an insoluble material. This process is well-known as an analytical technique for isolating certain radioactive materials. Water clarification processes make extensive use of this process by hydrolyzing and polymerizing Al ions to produce a gelatinous precipitate that co-precipitates other matter (e.g., ions, colloids, or even particles) [27].

After a LOCA, it is envisioned that various oxides, oxyhydroxides, and hydroxides of elements, such as Al, Zn, Fe, Si, and Ca, will be present in the system. These species have widely differing associated electrical charges and the relative charges on the different particles may change drastically with pH. It may therefore be necessary to characterize the surface charge characteristics of all components to determine whether mixed oxides, oxyhydroxides, and/or hydroxides can co-precipitate [27]. Radionuclides could also be co-precipitated through this mechanism.

There was no general consensus reached among the PIRT panelists on the importance of co-precipitation. Several panelists, citing a lack of evidence for co-precipitation in the ICET series, considered this issue to be less important than many of the other chemical effects previously discussed. However, at least one panelist believes that the ICET series may not have adequately simulated post-LOCA co-precipitation conditions, and that further research is warranted. Another panelist is concerned that the mass of iron(III) hydroxide ($Fe(OH)_3$) present will increase (e.g., from crud production, Section 5.3.2), and have a significant effect on coagulating and precipitating other materials. Research could be used to identify if either conservative or realistic post-LOCA environmental conditions are sufficient to foster co-precipitation and to characterize the species that form in concert with $Fe(OH)_3$ or other species. The research focus should be on identifying conditions that significantly alter either the precipitate concentration or the nature of the species that would form individually compared to other integrated [11] or single-effects testing [18].

5.4.2 Thermal Effects

Most of the previous discussion of chemical effects has focused on chemical processes that occur within, or are transported to, the containment pool. The containment pool water temperature (Table 13) is expected to peak between 190 °F and 270 °F soon (i.e., within 1 hour) after the hypothetical pipe break and then gradually decrease over the representative 30-day mission time for the ECCS recirculation system. After 15 days, the temperature range is expected to be between 115 °F and 165 °F.

Although the containment pool temperature will slowly decrease as the radioactive decay heat is removed from reactor fuel, ECCS water passing through the sump screen encounters two notable thermal transients during the residence time necessary for the water to complete a cycle through the ECCS cooling loop. First, the temperature decreases as coolant water passes through the residual heat removal (RHR) heat exchangers. The exact temperature decrease is a function of several variables (e.g., cooling water temperature, inlet water temperature, heat exchanger efficiency, and flow rates), but an approximately 30 °F decrease compared to the containment pool temperature is typical. Second, after passing through the heat exchangers, fluid is injected into the RCS and is heated as it is circulated through the RPV. The exact temperature increase within the reactor core will also depend on several variables (e.g., reactor and ECCS design and operability, break location, and time after the LOCA) and will diminish with distance from the fuel. However, a representative temperature adjacent to the reactor fuel is approximately 130°F above the containment pool temperature.

These two thermal transients may induce various chemical effects depending on the dissolved and solid chemical species entrained in the ECCS flow. Chemical effects and implications that impact the heat exchanger performance and heat removal from the reactor fuel are discussed separately in Sections 5.7.2 and 5.7.3.1. Additionally, reactor fuel deposition and spalling have been previously discussed in Section 5.3.2. Chemical effects related to solid species precipitation from the cooling thermal transient at the heat exchanger and the heating transient induced by the reactor fuel are characterized in this section.

The principal concern is that the thermal cycle existing within the ECCS coolant loop provides a mechanism for continued generation of solid species precipitation. Metallic corrosion and nonmetallic leaching could continue to generate dissolved ionic species which are then precipitated from the solution as solids by either the cooling or the heating transient. Under these conditions, precipitate formation will continue as long as complete precipitate redissolution at the containment pool temperature does not occur, and corrosion and/or dissolution continues to generate dissolved concentrations above the solubility limit of the precipitating species.

5.4.2.1 RHR Heat Exchanger (Cooling Transient) Effects

Chemical species having normal solubility profiles may be dissolved in the containment pool at higher temperatures. However, these chemical species may precipitate in the heat exchanger because of the approximately 30 °F temperature drop (T3-20, Table 9; T4-35, Table 10). Some possible solid species that could form include $Al(OH)_3$, $FeOOH$, and amorphous SiO_2[3]. The lower temperature at the heat exchanger outlet could also facilitate the development of macroscale coatings and/or suspended particulates. Single-species solubility relationships can provide some information about the possible precipitated species, but co-precipitation effects (Section 5.4.1) may also be an important consideration.

The primary implication of precipitation within the RHR heat exchangers discussed in this section is that they could contribute to head loss at the sump screen. Head loss would only result if the species remain insoluble at higher temperatures as they circulate through the ECCS loop through the reactor core and back into the containment pool, although some fraction of the ECCS flow will not pass through the core but go directly to

---

[3] Any colloidal silica that is formed will be changed to silicate because of radiolytic hydration within the reactor core.

the sump through the break. Precipitated materials from the heat exchanger will not have a chance to redissolve at higher temperatures for this fraction of ECCS flow. Additionally, the precipitates would need to be suspended within the ECCS flow so that they do not settle before reaching the sump screen.

The PIRT panelists generally agreed that solid species precipitation caused by the heat exchanger transient is a highly important phenomenon that could impact ECCS performance. This high ranking is based on the knowledge that changes in solubility (with temperature or pH) are a principal driving force promoting precipitation. Additionally, the solubility of those species that are most likely precipitated decreases with temperature. Some initial work has been conducted to investigate solubilities of species identified in ICET and single effects testing [14, 15, and 53]. However, co-precipitation and other effects have not been considered and more systematic solubility studies may be required to cover the spectrum of relevant chemical species and plant-specific variables.

5.4.2.2 Reactor Core (Heating Transient) Effects

The higher temperature within the RPV and, most importantly, at reactor core fuel cladding surfaces could cause materials with retrograde solubility (e.g., organics, Al, B, Ni, Fe, Zn, Ca, Mg, SiO_2, and CO_2-based deposits) to precipitate (T3-27, Table 9; T4-38, Table 10). Some of these precipitates may form within the ECCS coolant water away from the fuel cladding but the highest concentrations will occur adjacent to the fuel cladding itself. These precipitates could then bind to the reactor fuel surface through caking and self-cementation as discussed in Section 5.3.2. As with low-temperature precipitation, co-precipitation effects (Section 5.4.1) may also be an important consideration within the RPV.

One implication of precipitation within the RPV is that additional precipitate could be created and transported to the sump screen that would then contribute to head loss. Spalling of reactor fuel cladding deposits (Section 5.3.2.) could also create additional solid products which contribute to sump screen head loss. Precipitated or spalled materials passing through the sump screen may degrade the performance of ECCS components downstream from the screen as well (Section 5.7).

The majority of the PIRT panelists also identified solid species precipitation caused by the reactor core thermal transient as an important phenomenon that requires additional study. As with cooling transients, this phenomenon is highly ranked because changes in solubility (with temperature or pH) are a principal consideration leading to precipitation. Little NRC-sponsored or industry-sponsored research has been performed to date to evaluate the retrograde solubility of possible precipitated species.

5.4.3 PIRT Research Recommendations

Several issues related to the precipitation mechanisms discussed in this section could be clarified through additional study to determine either their generic or plant-specific relevance. The conditions leading to polymerization are a function of the ionic species and concentrations, time, temperature, and pH. Therefore, it will be necessary to either realistically or conservatively assess relevant combinations of these principal variables to determine if solid species precipitation and ripening through polymerization occurs under plant-specific conditions.

Such a study could also consider solid species precipitation caused by the cooling of the ECCS water at the heat exchanger and heating within the RPV. Associated research to evaluate effects of pH and temperature history variability, as recommended in Section 5.1.5, could be performed simultaneously. Investigation of retrograde solubility precipitation could also be combined with the research discussed in Section 5.3.5 to evaluate reactor fuel clad deposition and subsequent spallation.

As with the research described in Section 5.1.5, it is difficult to generically assess solid species precipitation caused by high- and low-temperature transients because of he expected differences in the post-LOCA environment among the PWR plants. Therefore, it may be necessary to assess plant-specific temperature, material, and other relevant environmental conditions (e.g., pH). It may be possible to design testing to either separately evaluate conditions at each plant or to conservatively bracket conditions at all plants or among groups of similar plants (i.e., those using the same chemical buffers).

Research could also be useful to identify if either conservative or realistic post-LOCA environmental conditions are sufficient to foster co-precipitation and to characterize the species that form. The initial research focus should be on identifying relevant conditions that could significantly alter either the precipitate concentration or the nature of the species that have been observed in single-effects [18] and integrated-effects testing [11]. Follow-on testing could then more systematically evaluate these conditions to determine if significant implications result from co-precipitation.

5.5 Solid Species Growth and Debris Transport

This section discusses several issues related to both growth and maturation of solid species precipitates (Section 5.4) and how these and other chemical phenomena affect debris transport. Issues related to debris transport include chemical implications for the transport of physical debris into and within the containment pool and the transport of chemical precipitates. Debris transportability is an important consideration because it ultimately governs the physical and chemical debris concentrations and the debris characteristics, which affect ECCS performance in recirculation mode. If it remains stable and intact, debris that does not transport to the sump screen may have a minimal impact on ECCS performance.

The pipe break that initiates the LOCA will create insulation, coating, corrosion products, and other physical debris (Section 5.3), which may be transported into the containment pool. Once containment sprays are initiated, additional quantities of this debris may be washed into the containment pool. Containment sprays may also wash additional latent debris, corrosion products, and failed coatings that were not directly created by the water jet into the containment pool. As previously discussed in Section 5.3.1, the panelists universally recognized that debris transported by the containment spray was an important contributor to the containment pool chemistry and associated chemical effect implications.

Other than creating debris, water emanating from the pipe break and containment sprays forms the pool in the bottom of the containment. The containment pool fills for approximately 20 minutes during a LB LOCA before water from the RWST is exhausted and the ECCS begins recirculating coolant water. During the time before recirculation, there is no directional flow within the containment pool toward the sump screens. In addition, the pool is expected to be relatively quiescent as it fills except for areas where the water from the pipe break flows into the pool (T2-11, Table 8). These areas are expected to exhibit substantial turbulence and mixing. Localized turbulence may also occur during this period near the surface as containment spray water impacts the containment pool surface.

Directional flow toward the sump screen begins once recirculation starts. This is the mechanism by which the physical and chemical debris that has been transported to, or created within, the containment pool can be transported to the sump screen. The debris will then either be filtered or will pass through the sump screen into the ECCS inlet piping. There are several chemical mechanisms that the PIRT panelists identified that may affect the transport of physical and chemical debris both within the containment pool to the sump screen or downstream of the sump screen if it is ingested. These mechanisms have been separated into particulate growth and agglomeration and other chemical and physical settling and deposition effects.

Current guidance [4] is generally conservative and sufficient to address the generation and transport of most physical debris (except for additional sources discussed in Section 5.3.2) into the containment pool. However, the transport of chemical products is not explicitly considered. As discussed previously (Section 5.3.1), pipe break scenarios that provide the greatest physical debris transport may not result in the most deleterious chemical effects. Therefore, it is important for a realistic chemical effect evaluation to consider transport scenarios that provide the greatest chemical effect challenges. Consideration of various transport scenarios may not be necessary for conservative evaluations that assume that all physical and chemical sources are transported to the sump screen and/or downstream of the sump screen.

5.5.1 Particulate Growth and Agglomeration

Precipitation is a process involving the formation of small submicroscopic or microscopic solid particulates. The impact and important variables affecting precipitation have been previously discussed (Section 5.4). However, the growth (ripening) of these particles through aging and/or their aggregation or coagulation to form larger clumps of particulates (agglomeration) determines the physical size and cohesiveness of the solid species (T2-21, Table 8; T3-14, Table 9). Conditions that promote ripening and/or agglomeration of precipitated solid species within the containment pool will lead to larger particulates that may be less vulnerable to transport (T4-40, Table 10). Although bigger particulates could decrease sump screen blockage if a large percentage settles, the larger particles that do transport are more likely to contribute to head loss across the sump screen. In addition, the larger particulates trapped on the sump screen will form a cake that is more effective at trapping smaller particles. As a result, head loss can increase with time instead of reaching steady-state.

The PIRT panelists typically ranked the effects of particulate aging and growth on transportability as being of moderate to high importance. Most panelists confirmed that particulate growth can be a beneficial phenomenon that would reduce screen blockage if particulate sizes that limit transportability are reached. One panelist commented that as the concentration of several species increases over time, the aging of insoluble materials should generally increase the size, density, and crystalline nature of these materials [47]. This process would be promoted throughout those portions of the containment pool characterized by either a relatively low flow rate or stagnant conditions. However, one panelist indicated that the effect is highly dependent on the plant design and pH buffer and may therefore be difficult to credit generically.

Agglomeration encompasses both inorganic and organic aggregation mechanisms. Inorganic agglomeration (T2-21, Table 8; T3-14, Table 9) is the process whereby colloidal particles (1–100 nm in diameter) that are formed by cation polymerization (Section 5.4.1) are electrically bound to create larger particle clusters and eventually visible precipitates. Organic agglomeration describes the ordering of organic colloidal particles (1–100 nm in diameter) into larger clusters. However, the presence of organic species can also either increase or decrease the likelihood of inorganic agglomeration. Organic acids or oil can promote inorganic agglomeration by binding smaller inorganic particulates into clusters (T2-24, Table 8). This mechanism is analogous to the manner in which soap coagulates dirt particles and is defined here as organic binding. Organic species can also discourage inorganic agglomeration through absorption (T2-15, Table 8). More discussion of organic agglomeration effects is contained in Section 5.6.2.

The importance of inorganic agglomeration was generally highly rated by the PIRT panelists. Inorganic agglomeration is sensitive to many variables including the pH of the point of zero charge (PZC) of the species, the ionic strength of the fluid, particle shape factors, and particle sizes. If a multi-component system contains particles of opposite charge, the particles will coagulate to form macroscopic precipitates. This electrostatic agglomeration occurs when the solution pH lies between the PZC of the different particles. Even if the particles are not oppositely charged, increasing the ionic strength of the medium will cause the electrical double layers of solid particulate to shrink to the extent that the attractive van der Waals forces will overcome electrostatic repulsion and coagulation will occur as in Al water clarification technology [65].

Most panelists agreed that inorganic agglomeration is most likely necessary to form particles large enough to degrade ECCS performance and that it could occur quickly if conditions were favorable. This stipulation implies that particulate growth and aging, by itself, may not be sufficient to degrade ECCS performance. At least one panelist believes that inorganic agglomeration contributed in the formation of the visible precipitates observed in ICET and other testing (Section 1.2), but does not require separate study.

5.5.2 Settling and Deposition

This section considers other important physical and chemical processes that affect the transportability of both chemical products and physical debris. Settling is defined as a lack of transport of either physical debris or chemical products within either the containment pool or the ECCS. Deposition refers to chemical products that are deposited on debris, component, or structural surfaces within the containment atmosphere, the containment pool, or the RCS. Deposition on the reactor fuel and within the heat exchanger is also discussed in Sections 5.3.2, 5.4.2, 5.7.2, and 5.7.3.

Both settling and deposition (T2-23, Table 8) can potentially remove substantial quantities of material that could otherwise be either trapped on the sump screen and contribute to head loss or affect downstream ECCS performance. Deposition can also inhibit metallic corrosion or nonmetallic leaching (Section 5.3) which would decrease the species present in the cooling water. Except for settling and deposition within the RPV or RHR heat exchangers (Sections 5.3.2, 5.4.2, 5.7.2, and 5.7.3), these are beneficial chemical phenomena. The challenge is in reliably determining how much credit can be given to settling and deposition in either generic or plant-specific post-LOCA conditions.

The containment pool directional flow rates during recirculation will be on the order of 0.05 ft/s (or less) in most existing PWR plants after the sump screens have been enlarged. This low flow rate should be prevalent through the bulk of the containment pool except for regions impacted by the flow of water through the pipe break. Obviously, low flow conditions are more conducive to settling of both chemical products and physical debris (T3-11, Table 9). Additionally, the containment floor is a maze of components bolted to the concrete floor such that the path of containment water from the break to the containment sump is less likely to involve linear, unidirectional flow. There are also drop out areas on the containment floor. These features provide trapping surfaces, flow stagnation regions, and longer flow paths, which all help retain chemical products within the containment pool. The PIRT panelists generally believe that the quiescent nature of the containment pool and the containment floor features are highly important and beneficial in reducing the impact of chemical effects.

Chemically induced settling (T3-13, Table 9) of solid debris can also be promoted by chemical species that attach to or coat the debris. Some examples provided by the PIRT panelists include the possibility of coating fiberglass fibers with Al to shift the PZC and enhance agglomeration, thus forming a hydrophobic organic coating on particulates, or forming an amorphous silica coating (T4-39, Table 10) on the surface of chemical precipitates (e.g., $Al(OH)_3$ and FeOOH). These effects have the added potential benefit that they could inhibit further chemical reactions for those materials. It was observed that Al species coated fiberglass and inhibited leaching in both ICET testing [11] and separate dissolution studies [17].

The average importance ranking was medium for chemically induced settling. One panelist commented that the amount of potential precipitate removed by precipitation on surfaces within containment is likely to be small in relation to material precipitating in suspension. However, the panelist also cautioned that the effect is dependent on pH and the buffering chemicals. Although one panelist thinks that amorphous silica will be precipitated given sufficient silicate sources (e.g., Cal-Sil), another panelist believes that flow through the core will produce radiolytic effects (Section 5.2) that will preclude its formation.

5.5.3 PIRT Research Recommendations

The PIRT identified agglomeration as an important mechanism for defining the characteristics, properties, and subsequent implications of precipitates that form during the post-LOCA scenarios. Therefore, it is necessary to understand those post-LOCA environmental variables that may affect agglomeration including solid species precipitate types and concentrations, the PZC of these species, and the ionic fluid strength. Variability in important post-LOCA parameters (e.g., materials, temperatures, pH, and buffering systems) should also be evaluated to determine if the effect on agglomeration is significant. Although it may be possible to conduct this evaluation generically, a plant-specific consideration should be more accurate given the variability in containment materials and conditions. This study may be particularly important for demonstrating that chemical precipitates in single-effects [18] or integrated-effects [11] laboratory testing are representative of products expected during a post-LOCA event. Similarly, it should be demonstrated that surrogate precipitates produced outside the testing environment [18] are representative of post-LOCA chemical species.

The transport phenomena discussed in this section generally benefit ECCS performance by removing physical debris and chemical products from circulation by settling or deposition[4]. Therefore, no research is necessary to assess their adverse implications. However, research may be necessary if these phenomena will be used to justify reducing the amount of suspended solid particulate in a chemical effects evaluation. It may be possible to study transport phenomena generically assuming that a parametric study can appropriately encompass the plant-specific variability in important transport variables such as plant design, containment materials, containment flow rates, pH buffer, and the chemical conditions resulting from appropriate hypothetical pipe break scenarios. Plant-specific analysis could also be used to decrease variability stemming from differences among plants.

A possible research strategy, if credit for settling is desired, could be to evaluate those precipitate characteristics (e.g., particle sizes, cluster sizes, and waters of hydration) under conditions that would least likely promote agglomeration. The conditions could be determined through testing or possibly analysis. Settling of the precipitate could then be evaluated under containment floor flow rates to determine the settling rate. Containment floor turbulence (e.g., under the break flow) should also be considered as a source to promote particulate suspension. Head loss from suspended solids could also be studied using the supernatant from above settling tests to evaluate cake formation and the rate of head loss across the sump screen.

The NRC-sponsored settling tests of $Ca_3(PO_4)_2$ provide an example of some of the issues associated with performing these tests. These tests were conducted in a large settling column [15]. The column was filled with a borated water solution containing TSP. Calcium chloride ($CaCl_2$) was introduced into the solution to form $Ca_3(PO_4)_2$. The solution was stirred to get a uniform mixture and then the precipitates were allowed to settle without further agitation. Two different $CaCl_2$ concentrations were tested to provide a higher (300 ppm (7×10^{-3} M)) and lower (75 ppm (1.9×10^{-3} M)) dissolved Ca inventory. In this simple test, the 300-ppm (7×10^{-3} M) concentration resulted in, on average, larger particles and faster settling times than the 75 ppm (1.9×10^{-3} M) concentration [15]. Thus, this simple test demonstrated the effect of precipitate concentration on both agglomeration and settling. However, one panelist was critical of the applicability of the results of that test because the conditions were not sufficiently representative of the post-LOCA containment pool. A particular concern was that the test flow rates were not matched with containment flow rates to ensure similar hydrodynamic forces.

5.6 Coatings and Organics

5.6.1 Coatings

Coatings existing within containment represent possible additional physical debris sources. Generally conservative guidance for considering the effects of physical coating debris is provided [4] for the evaluation of ECCS performance. However, both coatings and organic materials can impact the chemical effects that occur within, or are

---

[4] Possible adverse effects related to deposition are discussed in Sections 5.3.1, 5.4.2, and 5.7

transported to, the ECCS cooling water. Both inorganic (e.g., Zn-based) and organic (e.g., epoxy-based) coatings exist within containment. One concern is that these coatings leach chemicals as a result of being submerged in the containment pool environment after the LOCA (T3-29, Table 9; T4-17, Table 10). The coatings could either be intact on the substrate or loose within the containment pool as chips or smaller particulate. The implication is that coatings may create additional chemical species (e.g., chlorides or organics) within the containment pool. Some PIRT panelists mentioned that the most likely coatings that could leach significant quantities of deleterious materials include Pb-based paints (present in older containment buildings), phenolics, and vinyl coatings.

There has generally been little testing on leachant species and rates from nuclear-grade coatings [23]. The ICET series did incorporate inorganic Zn-coated steel to simulate material with a Zn-rich primer (T4-13, Table 10) that was either without a topcoat, or the topcoat had failed [12]. However, none of the previous NRC-sponsored or industry-sponsored testing (Section 1.2) has more broadly considered the chemical effect of coatings, principally because of the tremendous variability of types and quantities in nuclear plants.

The testing that has been done has evaluated only a few qualified epoxy and primer coatings and most often has examined only leachable Cl, F, sulfur (S) (or sulfate), and total halide contributions under a high-pH (i.e., 13), high-temperature (i.e., 95 °C), non-borated environment [22]. Chloride concentrations up to nearly 700 ppm (0.02 M) have been measured in this testing, but more typical concentrations fall between 100–200 ppm ($3-6 \times 10^{-3}$ M). Fluoride concentrations are generally less than 1 ppm (3×10^{-5} M) but have been measured as high as 100 ppm (3×10^{-3} M). The total organic halide concentration, when reported, is often around 1 ppm. The S or sulfate concentrations are often less than 100 ppm (3×10^{-3} M), but peak as high as 800 ppm (0.02 M) [22]. Note that the approximately scaled containment pool concentrations associated with these measured test concentrations can be determined by dividing the above concentrations by 50. One panelist notes that these leaching tests measure the Cl, F, and sulfate ionic concentrations. There are likely organically bound halogens and S in these coatings that mask the total element concentration.

Considering the general lack of applicable testing information in post-LOCA environments, the PIRT panelists appeared divided on the importance of the leaching of coatings on chemical effects. One panelist indicated that most coatings should be resistant to dissolution, and concentrations of any dissolved species from coating dissolution should also be minor. Therefore, in his opinion, the overall effect on the ECCS coolant composition and formation of secondary precipitates is likely to be small.

Another panelist indicated that this effect would be important only if various metallic-based (e.g., Pb) coatings are present. Alternatively, one panelist believes that although coatings may not lead to significant precipitate quantities, the decomposition of organics within coatings by thermolysis and radiolysis could affect the organic complexation (Section 5.6.2) and agglomeration (Sections 5.5.1 and 5.6.2) of other inorganic chemical products. Another panelist believes that leaching from coatings may substantially add to the ionic loading of the containment sump water if organically bound halogens and S are contained in these materials, which are then disbanded from their molecules by radiolysis. This panelist cited the degradation of the spent fuel pool resin into soluble sulfates that occurs within a few months at low doses (i.e., radians to kiloradians). This process will be accelerated by the higher radiation flux present in the post-LOCA environment. Other panelists indicated that these effects would be inherently plant-specific and could only be evaluated by testing specific qualified and unqualified coating systems to evaluate both chemical and radiolytic dissolution characteristics in post-LOCA environments.

5.6.2 Organics

The PIRT panelists generally indicated that organics could significantly affect the nature, properties, and amounts of chemical byproducts that form in the post-LOCA containment environment. The biggest uncertainties relate to the possible organic sources and concentrations that are located within containment and present in the containment pool after a break (T1-7, Table 7; T2-17, Table 8). Coatings, discussed above, and insulation debris (e.g., fiberglass and Cal-Sil) represent obvious organic sources, but other less obvious sources may also be present.

Organics stemming from fiberglass and Cal-Sil insulation were included to some extent in the ICET experiments. Both the fiberglass and Cal-Sil were heat treated to simulate temperature profiles existing in service and representatively bake off some of the organic resins. However, dissolved organic materials were likely present in the ICET solution based on the pigmented solution color (yellowish to rust colored) observed in each test [11]. The organics were attributed to decomposition products of the phenol formaldehyde resin-based polymer that coats the fiberglass insulation tested in the ICET program. The total organic content was measured in ICET Test 2. The baseline measurement at the onset of testing was 0.2 mg/L, increased to 7.3 mg/L by day 15, and then more slowly increased to 7.9 mg/L by the end of the test at day 30 [11].

Another possible organic source discussed in the PIRT is the reactor coolant pump (RCP) oil and small oil reservoirs for other pumps located inside the containment building. The scenario evaluated in the PIRT (T1-7 Table 7; T4-18, Table 10) considered the overflow, failure, or leakage of oil and other organics from either the RCP oil collection tanks or lube oil systems. This ingress could be caused by either a preexisting damage within the system or LOCA-induced damage. The oil collection tanks are more likely to fail, especially if the pipe break occurs in close proximity, because they are fabricated from approximately ½-inch carbon steel with epoxy coating. However, these tanks only typically contain a few gallons of residual oil. The RCP lube oil systems contain a bigger source of oil (approximately 250 gallons per pump), but these systems are less likely to fail unless there is leakage or direct LOCA impact.

The panelists agreed that a relatively large release of oil caused by an RCP oil system failure might significantly affect containment pool chemistry if oil concentrations on the order of parts-per-million are generated. The oil could alter metallic corrosion and/or nonmetallic dissolution rates and species (Section 5.3.3.2), precipitation and/or co-precipitation (Section 5.4.1), particle buoyancy, agglomeration and/or complexation (below). The tanks could also provide organic food sources for bacterial growth (Section 5.3.4). The potential for RCP oil reservoir damage or leakage during the LOCA should be considered for these reasons.

The three related organic effects that affect the size of solid particulate clusters include organic agglomeration (T3-14 Table 9; T2-24, Table 8), organic binding (T2-24, Table 8), and organic complexation (T2-15, Table 8). As with inorganic agglomeration (Section

5.5.1), these phenomena impact the transport properties of solid species precipitates (Section 5.5.2) and also the likelihood that particulates are captured at the sump screen, within the heat exchanger, the RPV, or in downstream locations. Particulate transport could also be affected by organic buoyancy (T4-15, Table 10).

As previously discussed (Section 5.5.1), organic agglomeration describes the ordering of organic colloidal particles (1–100 nm in diameter) into larger clusters. Organic binding describes the formation of smaller inorganic particulates into clusters surrounding organic acids or oils. The importance of both organic agglomeration and binding was also generally highly rated by the PIRT panelists, but there is more uncertainty surrounding this ranking. One panelist indicated that if organic materials act as glues to accrete and grow solid particles (or act as surfactants), then large coagulated particles may be formed. However, another panelist indicated that organic agglomeration depends greatly on the type and magnitude of organics existing within the containment pool. A third panelist expects the concentration of organics to be low. The large uncertainty surrounding this issue implies that additional consideration is likely necessary to determine the scope and applicability of these phenomena.

Conversely, organic complexing agents act to inhibit agglomeration. Aliphatic acids are well-known complexing agents and are a likely class of organic decomposition product within the post-LOCA environment. The action of these complexing agents can occur either by adsorption onto solid surfaces or by interaction in solution with metal ions. Organic surface complexation occurs if organic molecules (i.e., amines, acids, and heterocycles) adsorb on surfaces of ions or solids and inhibit the subsequent precipitation or growth of those species. Solid species may precipitate, but they remain relatively small in size (nano-scale) because of the inhibition caused by the adsorbed organics. These species are therefore less likely to agglomerate or grow to macroscopic sizes. Solute complexation effectively enhances solid species solubility limits. Precipitates that do form will grow more slowly and can grow to greater size although the quantity of total solids will be lower. Solute complexation could increase the transportability of these metal ions but decrease the likelihood that they are trapped at the sump screen or another location within the ECCS. In this regard, the implications of organic complexation are counter to those associated with organic agglomeration or binding.

The average importance ranking for organic complexation was medium, although there was disagreement among the panelists on the significance. Two panelists argued that a minimal amount of organic acid will exist in the post-LOCA containment pool and organic complexation will not be significant. However, two panelists indicated that radiolysis of organics could yield organic acids (or CO_2) and that these conditions have yet to be studied. Another panelist indicated that even minor amounts of organic acid can possibly have a significant effect on surface reactivity and hence complexation. This panelist indicated that the phenol formaldehyde resin-based insoluble polymer coating NUKON® fiberglass could dissolve into an alkaline aqueous phase that could inhibit solid species nucleation and growth. The uncertainty and disagreement surrounding this issue implies that additional consideration is warranted, especially if this phenomenon is credited for enhancing solubility limits of the expected precipitates.

Organic buoyancy describes a mechanism whereby organic materials coat existing precipitates (e.g., $Al(OH)_3$) or physical debris (e.g., Cal-Sil and fiberglass) to increase the buoyancy. This effect could enhance the transport of particulates or chemical byproducts to the sump screen. The PIRT panel, on average, rated this particular phenomenon to be of medium importance, although there was again significant uncertainty: two panelists provided high rankings and two others provided low rankings. The justification for the high rankings was that increased buoyancy has the potential to transport more solids to the sump screen without affecting particulate sizes. Therefore, larger particles, which are more likely to be trapped at the sump screen, could be transported than without the buoyancy effect.

The organics also could act as surfactants that could sorb onto the particle surfaces, cause the particles to become hydrophobic, and give the now-coated particles greater affinity for air bubbles, which could cause them to float as the coolant suspension becomes aerated by agitation. This phenomenon would be akin to flotation which is used in mineral milling to separate ore particles for further beneficiation or processing.

The low rankings were justified, in part, on the stipulation that increased buoyancy would most likely cause the particulates to float. If the sump screen is fully submerged, floating particulates would not affect head loss significantly, and they would be less likely to be ingested into the ECCS pump. Therefore, they would be much less detrimental than would suspended debris. One panelist also speculated that if these particles were ingested into the ECCS pump, they would be ground up by the impellers and would have little detrimental effect on downstream ECCS components. This issue may only have practical ramifications if any buoyancy increase restricts settling but does not cause particulates to float. Within this buoyancy range, this effect could suspend and promote transport of solids that would normally settle within the containment pool.

The last potentially significant organics and coating phenomenon for consideration is the hydrothermal hydrolysis of various organic and/or inorganic coating and insulation materials (T2-14, Table 8). One of the panelists [65] identified this phenomenon separately from the PIRT exercise. This panelist stipulated that hydrothermal hydrolysis could partially de-polymerize polymeric materials, producing materials ranging from small molecules to colloids. These colloids could subsequently aggregate to form gels. The potential practical impact on ECCS performance is that the de-polymerized materials could be more likely to transport and contribute to larger head loss at the sump screen than could the original product forms. At least one panelist doubts that this effect is significant. The elevated containment temperatures necessary for this phenomenon are relatively brief. Depolymerization would also require thermodynamic destabilization, which is unlikely in the temperature range under consideration. This reaction may only be possible within the reactor core. Additional evaluation of possible effects under either conservative or reasonably representative post-LOCA conditions would be required to assess this phenomenon.

5.6.3 PIRT Research Recommendations

There is a general lack of knowledge about the identity of leachant species and rates of release from nuclear grade coatings under the chemical and radiolytic conditions present in the post-LOCA containment environment. Generic coatings research is also complicated by the variety of coating systems present within PWR containments. Therefore, plant-specific evaluation of qualified and unqualified containment coating systems is likely to be a more effective approach, which will undoubtedly first require an accounting of the types and quantities of coatings present within containment. Potentially, the most significant issues to address, as indicated by the PIRT panelists, are leaching from metallic coatings and the decomposition of organics within coatings by thermolysis and radiolysis. The objective should be to determine the types and concentrations of the leachants (including organics) that are liberated under either realistic or conservative post-LOCA conditions.

More general consideration of the effects of organics first requires a realistic assessment of possible organic sources and concentrations that may exist in the containment pool. Insulation and coating debris are known to contain organics, but an accounting of other sources (e.g., oils and greases) is necessary to understand the concentrations and types of species. This may require plant-specific evaluation unless it can be demonstrated that concentrations are sufficiently low for the types of organics that may exist in containment.

If sufficient organic sources (e.g., electrical insulations and plastic coatings) are available, the influence of organic agglomeration and/or binding could be significant. Additional research would then be useful to determine the leaching behavior of either specific organics (if sources are limited) or samples from representative organic classes (if sources are prevalent) within the post-LOCA containment environment. An analogous sampling strategy was used in the Westinghouse Owners Group single-effects chemical testing [18]. Research should then examine the ability of the water-treated organic materials to coalesce and/or bind inorganic solids (e.g., thermal insulation materials) as they cool.

Organic complexation is largely an advantageous effect that will minimize ECCS degradation by effectively enhancing the solubility limits. Therefore, no research is recommended unless credit for this effect is desired. Organic buoyancy may also be beneficial unless the buoyancy increase is sufficient to restrict settling, but not so great to promote floatation. Research would be needed to determine if organic buoyancy or flotation can occur within the range of post-LOCA conditions existing at PWR plants and if the buoyancy suspends, but does not float, significant debris concentrations that would otherwise settle. If the effect can be demonstrated, it may be possible to calculate the suspended debris if the inorganic and organic particulates types and sizes are known. This effect becomes less important if the settling of containment pool solids is not credited in a chemical effects evaluation.

The implications of hydrothermal hydrolysis under either realistic or conservative post-LOCA conditions may also require some additional consideration. No previous NRC-sponsored or industry-sponsored research has evaluated this effect. A literature and theoretical scoping analysis could be initially used to define its importance. The next step would entail identifying susceptible organic and/or inorganic coatings and insulation materials by screening their physical and chemical properties. After screening, the most susceptible materials could be evaluated to determine the propensity for interactions among the various constituents produced by hydrothermal hydrolysis to ascertain whether even more resilient (i.e., less susceptible to thermal hydrolysis) byproducts form. Both interactions among hydrolysis byproducts and with other containment pool species should be addressed [65].

It is important to note that the reaction of the products of hydrothermal hydrolysis are often solids themselves and their reaction with other solids, particularly those having a zeta potential of opposite sign (equivalent to having PZC values that are on opposite sides of the pH of the medium), will be assisted by the electrostatic interaction between the particles. Furthermore, because the rates of these reactions are often controlled or significantly affected by mass transport, rheological properties, which tend to control the interpenetration of gel-like hydrolyzed material, are also significant.

5.7 Downstream Effects

Some possible chemical and biological products that may form within the RCS or ECCS have been discussed (Section 5.3). These products may increase the debris loading that is transported to the sump screen. Most of the phenomena in the preceding sections have considered possible contributions of chemical effects to the head loss associated with debris that is trapped by the sump screen. However, this debris could also pass through the sump screen, become ingested within the ECCS pump suction line, and potentially travel through the remainder of the ECCS cooling loop. New chemical products could also be created by thermal cooling and heating transients at the heat exchanger and near the reactor core, respectively (Section 5.4.2). Either ingested debris or newly created chemical products have the potential to degrade ECCS components downstream of the sump screen. These considerations are globally referred to as downstream effects.

The evaluation guidance supporting GL 2004-02 requests that downstream effects be addressed in the licensees' safety evaluations [4]. General industry approaches also exist for conducting the necessary evaluation within the RPV [66] and in all other downstream components and locations [67]. Much of the existing evaluation guidance, especially in [67], addresses the effects of physical debris but does not explicitly provide guidance for considering chemical effects. However, the NRC staff is currently reviewing [66] to ensure that potential chemical effects are appropriately considered.

The PIRT panel did consider how chemical effects could degrade performance in downstream components. The PIRT panelists believe that the ECCS pumps, the RHR heat exchangers, and the reactor core are the three major downstream components that are either the most likely to contribute to chemical effects or the most susceptible to chemical effects. The potentially significant chemical phenomena identified by the PIRT that are associated with each of these locations are discussed in this section. These PIRT results are intended to inform staff review [66].

5.7.1 ECCS Pumps

Debris that passes through the sump strainer screen is ingested into the ECCS pump suction leg and then passes through the pump. The PIRT identified potentially significant phenomena resulting from physical debris and chemically induced solids passing through the pump. The first group combines chemical phenomena that may promote pump seal abrasion or degradation. The second phenomenon analogously considers corrosion or erosion of the ECCS pump internals.

Pump seal abrasion (T3-19, Table 9; T4-33 Table 10) could result from abrasive wearing or scouring of the seals. Abrasion caused by physical debris is a prominent consideration in the downstream evaluation guidance [67]. However, chemical products can also contribute to abrasion, and the effect of chemical products is not addressed in the industry guidance [67]. Magnetite from crud generation and spallation from the RCS system (Section 5.3.2) was one example of a hard material cited by the panelists that could abrade the seals. Seal degradation could also result from leaching of the pump seal materials caused by immersion within the post-LOCA buffered containment environment. The implications associated with these phenomena are similar. Additional particles could directly contribute to head loss across the sump screen and/or deposition and clogging within the reactor core or heat exchanger. Dissolved species from these phenomena could also form additional solid precipitates, and the additional species could contribute to co-precipitation. Pump seal abrasion or degradation could also degrade pump performance, possibly to the point of inoperability.

The PIRT panelists indicated that the importance of these phenomena, on average, fell between medium and high. The panelists almost unanimously agreed that the impact of pump seal particulates or leachants on either the formation or promotion of additional solid species or degradation elsewhere in the ECCS would be small. Only a small amount of material would be generated if the entire seal disintegrated. The more prevalent concern is failure of the ECCS pump seals caused by crud or other chemical products. In general, the panelists believe that additional consideration may be warranted to consider this effect.

Corrosion or erosion of other pump metallic internals (T4-32, Table 10) is technically analogous to nonmetallic pump seal degradation or abrasion. The internals most susceptible to corrosion or erosion are tight tolerance components such as bearings, wear rings, and impellers. As with pump seal degradation, these components could be eroded by Fe_3O_4 particles from crud generation (Section 5.3.2) or corroded as a result of immersion within the post-LOCA buffered containment environment. The redox potential of the debris-laden fluid also affects the corrosion and/or erosion of these internal components.

The panelists' expectations of the implications of internal corrosion and/or erosion of these components were also similar to seal degradation. Panelists did not expect that particulate or dissolved species generated by pump internal corrosion and/or erosion would exist in sufficient quantities to significantly alter chemical solid species formation or otherwise degrade ECCS performance during the 30-day post-LOCA time period considered in the PIRT analysis. The most likely cited implication was that significant erosion or corrosion could ultimately lead to pump failure well beyond this 30-day time period. One panelist also indicated that any two-phase flow through the pumps would also enhance metallic erosion.

However, the average importance ranking of this phenomenon provided by the PIRT panel was high, greater than the ranking for seal degradation. The higher ranking is justified by the tight tolerance, and therefore higher speed, of fluid passing within and through these internal components and by the greater importance of these components to pump operability compared to pump seals. Therefore, the proper consideration of erosion and corrosion of pump internals may be important if chemical solid particulates are harder or more concentrated than the physical particulates that are evaluated as part of the downstream effects evaluation [67]. Any assessment of chemical effects will likely need to be plant specific given the variability in pump designs and materials and other relevant post-LOCA conditions.

5.7.2 RHR Heat Exchangers

After ECCS coolant water has passed through the ECCS pump, it is transported through several valves and piping until it reaches the RHR heat exchanger. The PIRT panel identified several chemical phenomena that could occur at the heat exchanger. Most effects stem from the approximately 35 °F to 40 °F temperature drop at the heat exchanger outlet. This is effectively the lowest temperature within the ECCS recirculation loop. The possibility of chemical precipitation (or co-precipitation) of species having decreasing solubility with decreasing temperature (T3-20, Table 9; T4-35, Table 10) was previously discussed in Sections 5.1.2, 5.4.1, and 5.4.2.1[5]. The implications pertaining to the additional solids inventory on NPSH at the sump screen were also discussed in these sections.

However, a few other phenomena were identified by the PIRT panel that could directly degrade heat exchanger performance. One phenomenon (T3-21, Table 9) is surface deposition on the heat exchanger tubes from a combination of previously formed debris (i.e., chemical precipitate and physical debris) that passed through the sump screen and newly precipitated chemical species resulting from the temperature drop at the heat exchanger. This concern is supported by the Cal-Sil and $Ca_3(PO_4)_2$ deposits that were observed in ICET Test 3 even without a significant temperature drop [11]. These deposits caused the impeller-type flow meters to fail on the eighth day of the 30-day test. Another related phenomenon identified by the PIRT is clogging near the inlet of the close-packed heat exchanger tubes from both physical debris and chemically induced solid species (T3-21, Table 9).

Both phenomena could cause decreased flow through the heat exchanger core and/or diminished heat transfer between the ECCS and heat exchanger cooling water. Diminished cooling of the ECCS water could ultimately decrease the capacity of the ECCS water to remove heat from the reactor core. The PIRT panel generally considered both these phenomena to be important as three of the five panelists provided a high ranking. Rationale supporting the high rankings included the argument that precipitates are most likely in the heat exchangers because of the temperature decrease from the inlet to the outlet. Subsequently, any precipitates that form may degrade the hydrodynamic and thermal performance of the heat exchanger.

However, at least one panelist who provided lower rankings agreed that deposition and clogging have been observed during long-term operation. However, the magnitude of the lost heat capacity will depend significantly on the type and properties of any precipitates. This panelist justified his lower ranking on the basis that heat exchanger performance would not degrade significantly during the first 24 hours because of the substantial design margins. Further, beyond 24 hours, the decay heat loading is significantly reduced so that the margins are even greater. These phenomena may require additional consideration of the margin and operating history associated with the heat exchangers to assess their significance. Industry guidance for evaluating heat exchanger performance is provided in [67].

5.7.3 Reactor Core

The fuel cladding surface is approximately 120 °F hotter than the sump pool temperature during the post-LOCA event. The function of the ECCS is to remove decay heat from the reactor fuel during a LOCA. In common LOCA scenarios, the ECCS cooling water is initially injected into the RPV through the cold leg of the reactor coolant loop to cool the

---

[5] Note that the precipitation would need to be rapid given the relatively short residence time at the RHR heat exchanger.

reactor fuel. The ECCS water flows through the inlet nozzle of the reactor core, upwards through the vessel core, and out of the vessel through the hot leg. The water travels around the reactor coolant loop and flows back into containment at the pipe break location. During latter phases of the LOCA, this flow path is typically reversed such that the cooling water is injected into the hot leg. Operators may continue to change flow from hot-leg to cold-leg injection based on core thermocouple readings.

As previously mentioned, the ECCS cooling water will contain physical debris, any solids developed by chemical processes elsewhere in the ECCS cooling system (e.g., containment pool and heat exchangers), and dissolved ions from corrosion, dissolution, erosion, or other processes. The PIRT panel identified several significant phenomena related to chemical phenomena and processes that could occur as this fluid passes through the RPV. The implications of these phenomena can be categorized as follows:

A. phenomena that directly affect the heat transfer properties of the fuel or the ECCS water,

B. phenomena that increase blockage and decrease coolant flow within the RPV, and

C. synergistic effects on the containment pool environment.

5.7.3.1 Effects on Fuel or ECCS Water Heat Transfer Properties

Precipitation of retrograde soluble chemical species (Section 5.4.2.2), deposition, and subsequent spalling of physical and chemical solids (Section 5.3.2) have previously been discussed in terms of their potential additional contribution to sump screen head loss. However, both of these mechanisms could also directly affect heat transfer from the reactor fuel (T3-24, Table 9). The effects of precipitation of solid phases with retrograde solubility (T3-27, Table 9; T4-37, Table 10) would increase the concentration of solids within the core. Subsequent deposition (T3-22, Table 9; T4-37, Table 10) of these and other previously produced chemical species could decrease the fuel's heat conduction values. After deposition, these solids could continue to build up and may potentially spall from the reactor fuel clad at some point.

The panelists raised the following three unique scenarios related to the propensity of deposits to spall and the possible implications:

(1) Deposits spall and add to the additional solid species inventory within the ECCS loop (Section 5.3.2) but spallation restores some of the original heat conduction properties of the fuel (T4-38, Table 10),

(2) Deposits spall and simultaneously remove cladding which releases fission products to containment (T4-24, Table 10), and

(3) Deposits remain on the reactor fuel cladding and may continue to build up, which decreases the heat transfer from the fuel.

Although three of the five PIRT panelists provided high importance rankings for fuel deposition, none of them speculated on which of these scenarios is more likely. However, a dissenting panelist believes that fuel deposition will not occur to any significant extent. There have been no studies of reactor fuel deposition as part of either NRC-sponsored or industry-sponsored research programs. Prior research has focused on evaluating chemical phenomena occurring at temperatures associated with the containment pool and have not addressed higher temperature phenomena that are representative of the reactor core. Therefore, additional consideration is necessary to determine if deposition can occur to any significant extent. If so, the implications, extent, and likelihood of each of these defined scenarios should also be evaluated.

Physical and chemical solid debris within the ECCS coolant water could also diminish the fluid's heat capacity and degrade the ability of the coolant to remove heat from the core. This phenomenon may be more severe at the bottom of the reactor because of the settling of particulate chemical products (T3-26, Table 9) and debris. Settling is more likely in the core because, for cold-leg injection, the upwards coolant flow rate is relatively low.

One example of a possible phenomenon in this category is borate precipitation caused by localized boiling at the fuel elements within the two-phase water/vapor region in the core. As the coolant water is vaporized, the B and possibly other species, remains. Borate precipitation is possible if concentrations increase above the solubility limit. These precipitates may subsequently adhere to the fuel elements and diminish heat transfer within the two-phase region. Alternatively, these precipitates may spall from the fuel and either travel back to the sump strainer and possibly contribute to head loss or settle within the core. Precipitates that settle may contribute to diminished heat transfer or blockage below the two-phase region of the core. Greater dissolved B concentrations are also expected toward the bottom of the core as a result of stratification from boiling and the relatively low coolant flow rate within the core. These higher B concentrations may promote additional B precipitation below the two-phase region of the core.

Three of the five PIRT panelists provided high importance rankings for these related phenomena. The high rankings were based on the adverse effect on core cooling and potential damage to the fuel. However, accurate evaluation of these related phenomena requires an understanding of the debris concentrations, types, and sizes in order to determine the heat transfer effects. One panelist commented that engineering evaluations may be sufficient to assess the impact on the core thermal and hydraulic transfer. An initial conservative or bounding analysis may be useful generically to determine if heat transfer remains sufficient. If this bounding analysis is not sufficient, additional plant-specific evaluation could be employed. Other mitigative strategies, such as reversing ECCS coolant injection flow (Section 5.7.3.2), might also be effective in addressing these phenomena.

5.7.3.2 Effects of Flow Blockages

Spalled fuel deposition products (Section 5.3.2) and precipitated retrograde soluble chemical species (Section 5.4.2.2), which do not deposit on the fuel, could also settle within the RPV. In one sense, settling within the vessel can be advantageous because the solids are sequestered and are not transported back to the sump screen where they could contribute to head loss. However, settling can also be potentially deleterious if flow passages to the fuel elements are either globally or locally impeded (T3-25, Table

9). Reduced flow within the RPV, if significant, has the potential to diminish heat transfer from the fuel.

Three out of five PIRT panelists again provided high importance rankings for these related phenomena. The rationale supporting the high rankings is that significant flow blockage could globally or locally degrade the ability of heat to be dissipated from the fuel elements, which might then lead to cladding failure. Justification for the low importance ranking was based on the likelihood of mitigating these effects through operator action. Current ECCS design also allows for hot-leg injection. Common plant operation calls for reversing flow after approximately 8 to 24 hours of a LOCA by switching from cold-leg to hot-leg injection [68]. Thereafter, flow injection can be periodically alternated between the hot and cold legs.

Flow reversal is used, in part, to mitigate concerns about borate precipitation caused by localized boiling at the fuel elements (Section 5.7.3.1). This action is intended both to partially back-flush the fuel element channels to clear debris that may be causing blockage and to redistribute and redissolve borate. Of course, flow reversal will also resuspend some percentage of any solids that have settled in the RPV back into the main flow. These solids might then be transported back to the containment pool.

Core flow blockage could also result from physical and chemical debris formed downstream of the RPV that is transported through the ECCS. Sources include debris that passed through the sump screen, solids formed within the heat exchanger (Section 5.4.2.2), and spalled deposits from elsewhere in the RCS piping (Section 5.3.2). These solids, during cold-leg injection, must first pass through the reactor core inlet nozzle and the lower plenum debris screen. The lower plenum debris screen possesses a larger mesh size than the sump screens, which reduces the likelihood of debris retention at the core inlet nozzle. However, solids may still be trapped at the screen or in other locations within the core (e.g., fuel grids, grid straps, and baffle holes) that serve as primary or secondary flow passages.

Suspended solids formed downstream of the RPV travel with the flow during cold-leg injection, whereas spalling or settling deposits from the fuel core travel against the prevailing upward flow. These different flow paths may result in preferential settling locations for each type of solid based on its physical and chemical characteristics. As before, the practical implication of either locally or globally restricting flow to the core is that heat transfer from the fuel could be diminished. The significance of this phenomenon was also ranked highly by three of the five panelists using similar rationale as that used to justify the importance of flow blockages from settling solid species (above).

These potential core blockage phenomena should be considered when evaluating ECCS performance. However, some initial evaluation of flow blockage at the reactor core inlet nozzle has been completed [69]. This study predicts a small, global temperature increase (approximately 5 °F) at the highest temperature fuel element only after the flow through the reactor core inlet nozzle is substantially blocked (approximately 95 percent of the flow area). This study did not address the effects of either a debris bed covering the inlet nozzle or localized fuel spacer blockages. Therefore, additional consideration of the impact of these effects on fuel element cooling is warranted.

5.7.3.3 Synergistic Effects on Containment Pool Environment

Water temperatures within the reactor core resulting from either elevated fuel temperatures or higher core inlet temperatures (caused by degraded heat exchanger efficiency) could impact chemical effects within the remainder of the ECCS. Higher core temperatures could increase or initiate precipitation of species with retrograde solubility (Section 5.4.2.2). Higher containment pool temperatures may also lead to more dissolution or corrosion of submerged materials, resulting in higher ionic concentrations of dissolved species [47]. Therefore, the chemical equilibrium initially established will be altered to yield different precipitate concentrations or possibly new precipitates that were not originally present [47]. The concentration of precipitates could increase with higher temperatures if the dissolution rate of extant materials increases more rapidly than the solubility of secondary precipitates or vice versa. If ECCS coolant temperatures do increase, the implications can be considered only by evaluating the solubility (Sections 5.1.2 and 5.4), retrograde solubility (Section 5.4.2.2) and corrosion and/or dissolution processes (Section°5.3.3.2) for plant-specific materials and conditions.

Another synergistic effect considered by the PIRT is related to H increases within the reactor core. Hydrogen concentration increases as cladding oxidation progresses (T3-23, Table 9). This effect is already a design-basis consideration as it must be demonstrated that hydride formation does not compromise fuel ductility [2] and that H deflagration is avoided in containment [70]. The unique aspect of this phenomenon raised in the PIRT is that additional H from this reaction could affect the redox potential and other chemical processes as previously discussed (Sections 5.1.4 and 5.2.1).

The significance of this phenomenon was ranked as low by three of the five PIRT panelists. The low ranking was justified by the rationale that this phenomenon will only liberate significant H early in the post-LOCA period as the hydrolysis rate decreases as the fuel is cooled. Additionally, any H_2 that forms will partition into the containment building atmosphere quickly where it is relatively inert. Therefore, it has minimal effect on the redox potential. The only high ranking was justified by the rationale that H within containment must be accounted for in order to accurately determine the redox potential and then evaluate the subsequent effects. Therefore, it is recommended that this source of H be considered when evaluating the redox potential of the containment pool water and applicable post-LOCA chemical effects.

5.7.4 PIRT Research Recommendations

The pump performance evaluation [67] already requires that the effects of debris-laden ECCS water on wear, corrosion, and erosion be considered. As plant-specific chemical products and concentrations are determined (Sections 5.1 and 5.3), the wear characteristics (i.e., hardness and particle size) should be assessed to determine if significant additional degradation results from these products. Other than Fe_3O_4 and other similar particles created during crud release (Section 5.3.2), most other chemical solids generated in the post-LOCA period are expected to be less detrimental than existing physical debris, particulates from insulation, coatings, and latent debris.

Demonstration of acceptable heat exchanger performance is also part of the downstream effects evaluation [67]. The only pertinent chemical effects related specifically to heat exchanger performance are deposition and clogging. It may be possible to evaluate depositional effects by considering that conservative deposit concentrations are uniformly distributed within the tubes to determine the reduction in heat removal compared to the requirements imposed by the LOCA scenario. Heat exchanger tube clogging is most likely at the heat exchanger tube inlet. Evaluation of the flow conditions at this location should provide information to evaluate the propensity for clogging. If the evaluations credit chemical species solubility as a function of temperature, it may be necessary to consider the formation of additional solids as the recirculating temperature decreases. Another important consideration for these evaluations is the time is takes for either limiting deposit thicknesses or flow obstructions to occur with postulated combinations of chemical and physical debris. These evaluations may benefit from the fact that deposition and clogging are expected to worsen over time, and heat exchanger margins increase with time as the temperature of the core decreases.

The impact of phenomena within the reactor core is the downstream area where additional fundamental study would be most beneficial because previous NRC-sponsored and industry-sponsored research has largely not addressed chemical effects at these high temperatures. Evaluation of the chemical products that precipitate and then deposit onto the reactor fuel (Section 5.4.2.2) could be used to determine deposit thicknesses and heat transfer characteristics. Spalling propensity could also be evaluated (Section 5.7.3) to first determine the additional solid products that could be formed and then assess their implications for either reactor core cooling (Section 5.7.3) or sump screen clogging (Section 5.3.2). Smaller scale modeling of the RCS using heated, nonradioactive fuel surrogates could be used to evaluate the significance of these phenomena. Any such evaluation should identify conditions necessary for spalling of the corrosion film on the reactor fuel, investigate settling propensity of these materials, and assess potential flocculating effects of these materials.

The effect of the physical and chemical solid debris, both formed within and transported to the reactor core, on the heat transfer capacity of the ECCS coolant water is another consideration that is expected to be evaluated [66]. Engineering evaluations may be sufficient to assess the impact on the core thermal and hydraulic transfer. An initial conservative or bounding analysis may be sufficient to generically determine if heat transfer remains acceptable. It may also be possible to credit mitigation strategies such as reactor flow reversal from hot-leg injection in this analysis. If a bounding analysis is not sufficient, plant-specific evaluation could be used to determine applicable margins.

There has been some generic consideration given to the global effects of partial flow blockage at the reactor core inlet nozzle [69]. However, the implications of uniform blockage, blockages in other locations, and localized heating effects within individual fuel elements have not been considered. Existing operator actions to reverse ECCS injection flow may also help mitigate flow blockages and could be credited if appropriately demonstrated. Alternatively, analysis could be used to determine whether there are locations within the core that are more susceptible to flow blockage effects. Those locations could then be analyzed using conservative assumptions (as in [69]) to determine the significance of these phenomena.

Finally, it will be important to consider potential synergistic effects related to reactor core temperature increases stemming from the phenomena discussed in this section when evaluating chemical precipitation within the containment pool (Sections 5.1.2 and 5.4), at the heat exchanger (Section 5.7.2), or within the reactor core (Section 5.7.3). As previously discussed (Section 5.2.1), H production within the core should be assessed when evaluating the redox potential of the containment pool water and applicable post-LOCA chemical effects.

6 CONCLUSIONS

The NRC convened an external peer review panel to both review the NRC-sponsored research conducted through the end of 2005 and to identify and evaluate additional chemical phenomena and issues that were either unresolved, or not considered in the original NRC-sponsored research. A PIRT exercise was conducted to support this evaluation in an attempt to fully explore the possible chemical effects that may affect ECCS performance during a hypothetical LOCA.

The PIRT process followed the general approach that has been used and refined over the last several years at the NRC. The following four technical evaluation criteria were established for ranking chemical phenomena most likely to do the following:

(1) contribute to sump screen clogging,
(2) affect downstream component performance,
(3) impact core heat transfer, or
(4) degrade structural integrity.

The panelists independently completed the PIRT evaluation to rank the significance and knowledge level associated with each phenomenon that was evaluated. These rankings were then finalized after discussing the rationales behind the independent rankings for those issues resulting in a strong difference of opinion among experts.

The PIRT results are not intended to represent a comprehensive set of chemical phenomena or issues within the post-LOCA environment. Instead, these phenomena should be considered in the context of important findings from past research (Section 1.2) and informed by research completed since the PIRT was conducted. It is anticipated that knowledge gained by related research will be considered along with the PIRT recommendations to identify and resolve existing knowledge gaps so that a more accurate review of chemical effects evaluation in the GL 2004-02 licensee submittals can be performed.

The significant PIRT issues that were identified are grouped topically in the report. Significant issues were identified by the PIRT panel pertaining to the underlying containment pool chemistry; radiological considerations; physical, chemical, and biological debris sources; solid species precipitation; solid species growth and transport; organics and coatings; and downstream effects. At the end of each topical area discussion, a summary of issues that could benefit from additional evaluation is provided.

Several of these issues can be addressed using existing knowledge associated with the chemical effect. Additional analysis may only be needed to assess the implications of these effects over the range of existing generic or plant-specific post-LOCA conditions. Other issues may require additional study to more fully understand the chemical effects and determine if they are relevant. Then, the practical generic or plant-specific implications can be assessed. These issues are summarized below for each topical area.

6.1 Underlying Containment Pool Chemistry

- Additional investigation on the effect of concentrated boron salts within the RPV could determine whether any significant corrosion of the reactor vessel internals and fuel cladding occurs in the post-LOCA period.

- Also, the effects of concentrated borate on the types, amounts, and properties of chemical precipitates that could form within the RPV should be considered.

- Plant-specific pH and temperature history should be considered in combination with various postulated debris mixtures within the buffered environment. This evaluation would provide a necessary foundation for accurately determining the dissolved species, material corrosion and/or dissolution rates, and precipitate formation based on solubility and kinetic considerations. Variability in important factors should also be evaluated to determine the effect of changes or uncertainty in the underlying environment.

- Additional scoping calculations are needed to assess the magnitude and relative effect of H production from a variety of sources, including the Schikorr reaction, to determine if dissolved H plays a significant role.

6.2 Radiological Considerations

- Consideration of radiological effects is warranted to first understand the magnitude of the concentrated radioactive fields in the post-LOCA environment. Then, the effects of changes in the redox potential on the containment pool chemistry, material corrosion rates, and chemical speciation and solubility can be evaluated.

- The likelihood of transport and accumulation of activated species (Section 6.3) at the sump screen debris bed should be assessed to determine the strength of the radiation field. Localized chemical effects resulting from radionuclides trapped within the debris bed and the effect of radiation on the properties of precipitated solid species should then be considered.

6.3 Physical, Chemical, and Biological Debris Sources

- The possibility and extent of corrosion product spallation from reactor vessel walls, primary system piping and/or components, and fuel cladding surfaces under conditions that would occur under post-LOCA thermal and mechanical shock should be examined.

- The possibility that potential chemical precipitates could form as a result of retrograde solubility at the hot reactor fuel clad surface and subsequently spall should be investigated.

- Plant-specific debris sources, dissolved Ca concentrations, and sump environmental conditions should be considered to assess the implications of

$CaCO_3$ formation from atmospheric CO_2 on both the solid species inventory and, more importantly, its effect on precipitation kinetics. However, the amount of CO_2 provided by the air in the containment vessel may produce only negligible calcite compared to containment pool debris generated from other sources (e.g., fiberglass insulation, concrete, and Cal-Sil).

- Realistic, plant-specific containment pool environment, materials, and concentrations should be considered to identify the appropriate corrosion acceleration or inhibition factors. One acceleration factor that should be studied is the catalytic effect of possible containment materials such as copper and citrate (or potentially other hydroxy-organic acid anions, if present).

- Possible corrosion inhibition effects include concrete and other nonmetallic material aging products. Specifically, phosphate, chromate, dichromate, silicate, and borate can inhibit Al corrosion and deposition of Cu on Al may inhibit or accelerate corrosion. It will be important to understand plant-specific conditions in order to credit these inhibition effects.

- The effect of significant alloying additions in nuclear structural materials and debris or compositional changes in nonmetallic materials within the containment on the material dissolution rates should be considered. This evaluation should also determine whether significant dissolved concentrations of alloying or secondary elements are present that could alter chemical effects.

- Galvanic couples that may exist within the containment pool as a result of direct contact between dissimilar metals should be identified and evaluated to determine if they alter material corrosion rates.

- The viability and potential consequences of biological growth in a post-LOCA borated-water environment should be investigated. If significant growth is possible within 30 days, an understanding of important post-LOCA variables that may accelerate or retard the biological growth rate would be necessary to both identify locations within the ECCS where growth is possible and evaluate the associated implications.

6.4 Solid Species Precipitation

- A realistic or conservative assessment of relevant combinations of time, temperature, pH, and ionic strength should be performed to determine the types, sizes, and concentrations of solid precipitates that may form under plant-specific conditions and the significance of ripening through polymerization on these precipitates.

- Precipitation studies should also consider solid species precipitation caused by the cooling of the ECCS water at the heat exchanger and heating within the RPV (Section 6.3).

- Research could also be useful to identify whether conservative or realistic post-LOCA environmental conditions are sufficient to foster co-precipitation and to characterize any species that form.

6.5 Solid Species Growth and Debris Transport

- Variability in important post-LOCA variables (e.g., materials, temperatures, pH, and buffering systems) should also be evaluated to determine if the effect on agglomeration is significant. This study could be used to demonstrate that both chemical precipitates created in single-effects or integrated-effects testing and surrogate precipitates produced outside the testing environment are representative of materials that form in post-LOCA conditions.

- Research on transport phenomena may be necessary to validate reducing the suspended solid particulate caused by settling and deposition. Variability in the plant design, containment materials, containment flow rates and turbulence, pH buffer, and the chemical conditions present resulting from various hypothetical pipe breaks are important considerations that may affect the settling rates of containment pool debris.

6.6 Organics and Coatings

- The potentially significant coating issues to address, as indicated by the PIRT panelists, are leaching from metallic coatings and the decomposition of organics within coatings by thermolysis and radiolysis under either realistic or conservative post-LOCA conditions.

- More general research in the effects of organics should be preceded by a realistic assessment of possible post-LOCA organic sources (e.g., oils, greases, electrical insulations, plastic coatings, and paints) and concentrations in the containment pool. If sufficient organic sources exist, additional research would then be useful to determine the leaching behavior of these sources and examine the ability of the organic materials to coalesce and/or bind inorganic solids as they cool within the post-LOCA containment pool and consequently affect flow through the sump strainer screen.

- The interaction between organic materials and inorganic solids may increase the buoyancy of inorganic solids such that it is sufficient to restrict settling but not so great as to promote flotation. Additional study would need to determine if organic buoyancy can occur within the range of post-LOCA conditions existing at PWR plants and if the effect suspends, but does not float, significant debris concentrations that would otherwise settle. The engineering impact of organic buoyancy is significantly diminished if settling of containment pool solids is not credited in a chemical effects evaluation.

- The implications of hydrothermal hydrolysis under either realistic or conservative post-LOCA conditions have not been studied. The most susceptible coatings and insulation materials could be evaluated to determine the propensity for interactions among hydrothermal hydrolysis byproducts and with other containment pool species.

6.7 Downstream Effects

- The corrosion and erosion characteristics (i.e., hardness and particle size) of chemical products and concentrations could be assessed to determine if pump seal and internal components can be adversely affected. Magnetite and other similar particles created during crud release are expected to potentially be the most detrimental.

- The pertinent chemical effects related specifically to heat exchanger performance are deposition and clogging. A conservative analysis of deposition may be sufficient to determine if adequate heat exchanger design margins remain. The precipitation kinetics could be studied to determine if solid species concentrations at these lower temperatures are less than those assumed from either aqueous concentrations or solubility considerations. Evaluation of the flow conditions at the tube inlet location may also provide information to evaluate the propensity for heat exchanger clogging.

- Evaluation of the chemical products that precipitate and subsequently adhere to the reactor fuel (Sections 6.3 and 6.4) could be useful to determine deposit thickness and heat transfer characteristics. The propensity for spallation of chemical products from the fuel cladding could also be studied to evaluate the additional solid products that may be formed and their implications for either reactor core or sump screen clogging.

- The effect of the physical and chemical solid debris, both formed within and transported to the reactor core, on the heat capacity of the ECCS coolant water should also be considered. However, it is expected that the debris loading would have to be significant before the heat capacity is affected. Therefore, conservative engineering evaluations may be sufficient to assess the impact on the core thermal and hydraulic transfer.

- Implications of flow blockages in locations other than the reactor core inlet nozzle and localized heating effects within individual fuel elements should also be considered. It may be possible to use analysis or to credit existing operator actions to demonstrate that acceptable fuel heat transfer is maintained.

- It will be important to be mindful of potential synergistic effects related to reactor core temperature increases stemming from the downstream effect phenomena (Section 5.7.3.3) when evaluating chemical precipitation within the containment pool, at the heat exchanger, or within the reactor core.

- Hydrogen, H_2O_2, O_2, and OH^- production within the reactor core should be considered when evaluating the redox potential of the ECCS water (Section 6.2) and other applicable post-LOCA chemical effects.

7 REFERENCES

[1] R. Emrit, R. Riggs, W. Milstead, J. Pittman, and H. Vandermolen, "A Prioritization of Generic Safety Issues," NUREG-0933, U.S. Nuclear Regulatory Commission (NRC), Washington, DC, October 2006.

[2] *U.S. Code of Federal Regulations*, "Acceptance Criteria for Emergency Core Cooling Systems for Light-Water Nuclear Power Reactors," Section 50.46, Title 10.

[3] Generic Letter 2004-02, "Potential Impact of Debris Blockage on Emergency Recirculation during Design-Basis Accidents at Pressurized-Water Reactors," September 13, 2004.

[4] Safety evaluation by the Office of Nuclear Reactor Regulation related to the NRC Generic Letter 2004-02, Nuclear Energy Institute (NEI) guidance report (proposed document number NEI 04-07), "Pressurized Water Reactor Sump Performance Evaluation Methodology," Revision 0, December 6, 2004.

[5] A. R. Pietrangelo (NEI), letter to J. H. Hannon (NRC), "Pressurized Water Reactor Sump Performance Evaluation Methodology," (proposed document number NEI 04-07), May 28, 2004 (Agencywide Documents Access and Management System (ADAMS) Accession No. ML041550279).

[6] W.H. Ruland (NRC), letter to A.R Pietrangelo (NEI), "Draft Guidance for Review of Final Licensee Responses to Generic Letter 2004-02, 'Potential Impact of Debris Blockage on Emergency Recirculation During Design Basis Accidents at Pressurized-Water Reactors,'" September 27, 2007 (ADAMS Accession No. ML072600337).

[7] M. Bonaca (Advisory Committee for Reactor Safeguards), memorandum to W.D. Travers (NRC Executive Director for Operations), "Proposed Resolution of GSI–191, Assessment of Debris Accumulation on PWR Sump Pump Performance," February 20, 2003 (ADAMS Accession No. ML030520101.

[8] W.D. Shults (Oak Ridge National Laboratory), Analytical Chemistry Report to J.A. Daniel (General Public Utilities Service Corporation), September 14, 1979.

[9] Electric Power Research Institute, Nuclear Safety Analysis Center, "Analysis of Three Mile Island—Unit 2 Accident," NSAC-1, March 1980.

[10] B.C. Letellier, "Small-Scale Experiment: Effects of Chemical Reactions on Debris-Bed Head Loss," NUREG/CR-6868, U.S. Nuclear Regulatory Commission, Washington, DC, March 2005 (ADAMS Accession No. ML050900260).

[11] J. Dallman, B.C. Letellier, J. Garcia, J. Madrid, W. Roesch, D. Chen, K. Howe, L. Archuleta, and F. Sciacca, "Integrated Chemical Effects Test Project: Consolidated Data Report," NUREG/CR-6914, Volumes 1–6, U.S. Nuclear Regulatory Commission, Washington, DC, September 2006.

[12] T.S. Andreychek, "Test Plan: Characterization of Chemical and Corrosion Effects Potentially Occurring Inside a PWR Containment Following a LOCA," Revision 13, Westinghouse Electric Company, Monroeville, PA, July 2005 (ADAMS Accession No. ML052100426).

[13] V. Jain, X. He, and Y.M. Pan, "Corrosion Rate Measurements and Chemical Speciation of Corrosion Products Using Thermodynamic Modeling of Debris Components To Support GSI-191," NUREG/CR-6873, U.S. Nuclear Regulatory Commission, Washington, DC, April 2005 (ADAMS Accession No. ML051610123).

[14] M. Klasky, J. Zhang, M. Ding, B.C. Letellier, D. Chen, and K. Howe, "Aluminum Chemistry in Prototypical Post-LOCA PWR Containment Environment," NUREG/CR-6915, U.S. Nuclear Regulatory Commission, Washington, DC, October 2006 (ADAMS Accession No. ML062400465).

[15] K. Kasza, J.H. Park, B. Fisher, J. Oras, K. Natesan, and W.J. Shack, "Chemical Effects Head-Loss Research in Support of Generic Safety Issue 191," NUREG/CR-6913, U.S. Nuclear Regulatory Commission, Washington, DC, September 2006 (ADAMS Accession No. ML062280415).

[16] C.B. Banh, K.E. Kasza, and W.J. Shack, "Technical Letter Report on Follow-on Studies in Chemical Effects Head-Loss Research; Studies on WCAP Surrogates and Sodium Tetraborate Solutions," Argonne National Laboratory, Argonne, IL, February 15, 2007 (ADAMS Accession No. ML070580086).

[17] J. McMurry, V. Jain, X. He, D. Pickett, R. Pabalan, and Y.M. Pan, "GSI-191, PWR Sump Screen Blockage Chemical Effects Tests-Thermodynamic Simulations," NUREG/CR-6912, U.S. Nuclear Regulatory Commission, Washington, DC, September 2006 (ADAMS Accession No. ML060230442).

[18] A.E. Lane, T.S. Andreychek, W.A. Byers, R.J. Jacko, E.J. Lahoda, and R.D. Reid, "Evaluation of Post-Accident Chemical Effects in Containment Sump Fluids to Support GSI-191," Westinghouse Commercial Atomic Power (WCAP)-6530-NP, Revision 0, Westinghouse Electric Company, Monroeville, PA, February 2006 (ADAMS Accession No. ML060890509).

[19] Final safety evaluation by the NRR, "WCAP-16530-NP, Final Safety Evaluation of Post Accident Chemical Effects in Containment Sump Fluids to Support GSI-191," Pressurized Water Reactor (PWR) Owners Group Project No. 694, December 21, 2007 (ADAMS Accession No. ML073520891).

[20] J. McMurry and X. He, "Supplementary Leaching Tests of Insulation and Concrete for GSI-191 Chemical Effects Program," Technical Letter Report

IM 20.12130.01.001.320, Center for Nuclear Waste Regulatory Analyses, San Antonio, TX, November 2006 (ADAMS Accession No. ML063330573).

[21] American National Standard N101.2, "Protective Coatings (Paints) for Light Water Nuclear Reactor Containment Facilities," American Institute of Chemical Engineers, New York, New York, NY, 1972.

[22] F.P. Schiffley, II letter to the NRC, PWR Owners Group responses to the NRC second request for additional information on WCAP-16530, "Evaluation of Chemical Effects in Containment Sump Fluids to Support GSI-191," OG-07-129, April 3, 2007 (ADAMS Accession No. ML070950119).

[23] K. Natesan and R. Natarajan, "Survey on Leaching of Coatings Used in Nuclear Power Plants: Letter Report," Argonne National Laboratory, Argonne, IL, September 13, 2006 (ADAMS Accession No. ML062560368).

[24] American National Standard N5.12, "Protective Coatings (Paints) for the Nuclear Industry," American Institute of Chemical Engineers, New York, NY, 1974.

[25] R.L. Sindelar, M.E. Dupont, N.C. Lyer, P.S. Lam, T.E. Siddmore, P.R. Utsch, and P.E. Zapp, "Degradation and Failure Characteristics of NPP Containment Protective Coating Systems—Interim Report," Westinghouse Savannah River Company Topical Report (WSRC-TR)-2000-00079, March 2000.

[26] Electric Power Research Institute Report No. 1011753, "Design Basis Accident Testing of Pressurized Water Reactor Unqualified Original Equipment Manufacturer Coatings," September 2005 (ADAMS Accession No. ML071130069).

[27] P.A. Torres, "Peer Review of GSI-191 Chemical Effects Testing Program," NUREG-1861, U.S. Nuclear Regulatory Commission, Washington, DC, December 2006 (ADAMS Accession No. ML063630498).

[28] B.E. Boyak, R. Duffy, P. Griffith, G. Lellouche, S. Levy, U. Rohatgi, G. Wilson, W Wulff, and N. Zuber, "Quantifying Reactor Safety Margins: Application of Code Scaling, Applicability, and Uncertainty Evaluation Methodology to a Large-Break, Loss-of-Coolant Accident," NUREG/CR-5249, U.S. Nuclear Regulatory Commission, Washington, DC, December 1989.

[29] B.E. Boyack, A.T. Motta, K.L. Peddicord, C.A. Alexander, R.C. Deveney, B.M. Dunn, T. Fuketa, K.E. Higar, L.E. Hochreiter, S. Langenbuch, F.J. Moody, M.E. Nissley, J. Papin, G. Potts, D.W. Pruitt, J. Rashid, D.H. Risher, R.J. Rohrer, J.S. Tulenko, K. Valtonen, N. Waeckel, and W. Wiesenack, "Phenomenon Identification and Ranking Tables (PIRTs) for Rod Ejection Accidents in Pressurized Water Reactors Containing High Burnup Fuel," NUREG/CR-6742, U.S. Nuclear Regulatory Commission, Washington, DC, September 2001.

[30] S.M. Bajorek, A. Ginsberg, D.J. Shimeck, K. Ohkawa, M.Y. Young, L.E. Hochreiter, P. Griffin, Y. Hassan, T. Fernandez, and D. Speyer, "Small

Break Loss of Coolant Accident Phenomena Identification and Ranking Table (PIRT) for Westinghouse Pressurized Water Reactors", Ninth International Topical Meeting on Nuclear Reactor Thermal Hydraulics (NURETH-9), San Francisco, CA, October 3–8, 1999.

[31] R.K. Ratnayake, L.E. Hochreiter, S. Ergun, A.J. Baratta, "Identification and Ranking of Phenomena Leading to Peak Cladding Temperatures in Boiling Water Reactors during Large-Break Loss of Coolant Accident Transients," *Proceedings of the 10th International Conference on Nuclear Engineering (ICONE 10)*, Paper No. ICONE 10–22288, American Society of Mechanical Engineers, New York, NY, April 2002.

[32] J.R. Cavallo, T.S. Andreychek, J. Bostelman, B. Boyack, G. Dolderer, and D. Long, "PWR Coatings Research Program Phenomena Identification and Ranking Tables," Industry Coatings PIRT Report No. IC 99-02, June 16, 2000 (ADAMS Accession No. ML003733475).

[33] R.N. Morris, D.A. Petti, D.A. Powers, and B.E. Boyack, "TRISO-Coated Particle Fuel Phenomenon Identification and Ranking Tables (PIRTs) for Fission Product Transport Due to Manufacturing, Operations, and Accidents," NUREG/CR-6844, Volume 1, U.S. Nuclear Regulatory Commission, Washington, DC, July 2004.

[34] P.L. Andresen, F.P. Ford, K. Gott, R.L. Jones, P.M. Scott, T. Shoji, R.W. Staehle, and R.L. Tapping, "Expert Panel Report on Proactive Materials Degradation Assessment," NUREG/CR-6923, U.S. Nuclear Regulatory Commission, Washington, DC, February 2007.

[35] B.E. Boyack, T.S. Andreychek, P. Griffith, F. E. Haskin, and J. Tills, "PWR Debris Transport in Dry Ambient Containments-Phenomena Identification and Ranking Tables (PIRTs)," LA-UR-99-3371, Revision 1, Los Alamos National Laboratory, Los Alamos, NM, July 1999.

[36] B.E. Boyack, T.S. Andreychek, P. Griffith, F.E. Haskin, and J. Tills, "PWR Debris Transport in Ice Condenser Containments- Phenomena Identification and Ranking Tables (PIRTs)", LA-UR-99-5111, Revision 1, Los Alamos National Laboratory, Los Alamos, NM, December 1999.

[37] G.E. Wilson and B. E. Boyack, "The Role of the PIRT Process in Experiments, Code Development, and Code Applications Associated with Reactor Safety Analysis," Nuclear Engineering and Design, Volume 186, pp. 23–37, 1998.

[38] A.T. Howell (NRC), memorandum to W.F. Kane (RC), "Degradation of the Davis-Besse Nuclear Power Station Reactor Pressure Vessel Head Lessons Learned Report," September 30, 2002 (ADAMS Accession No. ML022740211).

[39] W. Arcieri, R. Beaton, D. Fletcher, and J. Lehning, "Calculation of Sump Fluid Temperature During Loss of Coolant Accidents at a Sample Nuclear Power Plant," Revision 1, Information Systems Laboratories, Inc., Rockville, MD, May 2006 (ADAMS Accession No. ML061510478).

[40] R.D. Reid, K.R. Crytzer, A.E. Lane, and T.S. Andreychek, "Evaluation of Alternative Emergency Core Cooling System Buffering Agents," WCAP-16596-NP, Westinghouse Electric Company LLC, Pittsburgh, PA, July 2006. This report is available as Attachment 1, "Response to Request for Additional Information Related to the License Amendment Request on Change of Containment Building Sump Buffering Agent from Trisodium Phosphate to Sodium Tetraborate," letter from J.A. Reinhart to the NRC, September 6, 2006, (ADAMS Accession No. ML062570173).

[41] D.D. Macdonald and M. Urquidi-Macdonald, "The Electrochemistry of Nuclear Reactor Coolant Circuits, "*Encyclopedia of Electrochemistry*, (A.J. Bard and M. Stratmann eds.), Volume 5, *Electrochemical Engineering*, (D.D. Macdonald and P. Schmuki, eds.), Wiley-VCH Verlag GmbH & Co. KGaA, Weinheim, Germany, pp. 665–720, 2007.

[42] J.R. Divine, W.M. Bowen, D.B. Mackey, D.J. Bates, and K.H. Pool, "Prediction Equations for Corrosion Rates of A-537 and A-516 Steels in Double Shell Slurry, Future PUREX, and Hanford Facilities Wastes," PNL-5488, Pacific Northwest National Laboratory, Richland, WA, 1985.

[43] S.B. Clark and C.H. Delegard. "Plutonium in Concentrated Solutions," Chapter 7 in *Advances in Plutonium Chemistry -- 1967-2000*, (D. Hoffman, ed.), American Nuclear Society, La Grange Park, IL, 2002.

[44] N. Wiberg, *Inorganic Chemistry*, Academic Press, San Diego, CA, 2001, p. 980.

[45] U.S. Nuclear Regulatory Commission, "Alternative Radiological Source Terms for Evaluating Design Basis Accidents at Nuclear Power Reactors," Regulatory Guide 1.183.

[46] E.C. Beahm, R.A. Lorenz, and C.F. Weber, "Iodine Evolution and pH Control," NUREG/CR-5950, U.S. Nuclear Regulatory Commission, Washington, DC, December 1992.

[47] R. Litman, "Evaluation of Integrated Chemical Effects Testing," Appendix D to P.A. Torres, "Peer Review of GSI-191 Chemical Effects Testing Program," NUREG-1861, U.S. Nuclear Regulatory Commission, Washington, DC, December 2006 (ADAMS Accession No. ML063630498).

[48] W.L. Lindsay, *Chemical Equilibria in Soils*, Wiley Interscience, New York, NY, 1979.

[49] J.P. Burke (NRC), letter to A.L. Hiser, Jr. (NRC), "Review of Southwest Research Institute's Technical Letter Report for Contract NRC-03-07-046," June 26, 2008, (ADAMS Accession No. ML081780515).

[50] C.J. Shaffer, D.V. Rao, M.T. Leonard, and K.W. Ross, "Knowledge Base for the Effect of Debris on Pressurized Water Reactor Emergency Core Cooling Sump

Performance," NUREG/CR-6808, U.S. Nuclear Regulatory Commission, Washington, DC, February 2003 (ADAMS Accession Nos. ML030780733 and ML030920540).

[51] C.W. Enderlin, B.E. Wells, M. White, F. Nigl, A. Guzman, D.R. Rector, T.J. Peters, and E.S. Mast, "Experimental Measurements of Pressure Drop Across Sump Screen Debris Beds in Support of Generic Safety Issue 191," NUREG/CR-6917, U.S. Nuclear Regulatory Commission, Washington, DC, January 2007.

[52] NRC Bulletin No. 2003-01, "Potential Impact of Debris Blockage on Emergency Sump Recirculation at Pressurized-Water Reactors," U.S. Nuclear Regulatory Commission, Washington DC, June 9, 2003 (ADAMS Accession No. ML031600259).

[53] R. Reid, "Incorporation of Additional Plant Inputs in the Chemical Effects Spreadsheet PA-SEE-0354," PWR Owners Group Presentation at the public meeting with NEI, PWR Owners Group, licensees, and sump strainer vendors to discuss the resolution of Generic Safety Issue 191, April 18, 2007. Meeting information can be accessed at http://www.nrc.gov/reactors/operating/ops-experience/pwr-sump-performance/public-mtgs/2007/index.html.

[54] F. Dacquait, C. Andrieu, M. Berger, J-L Bretelle, and A. Rocher, "Corrosion Product Transfer in French PWRs During Shutdown," Société Française d'Energie Nucléaire (SFEN)-Chimie Conference on Water Chemistry in Nuclear Reactor Systems, Avignon, France, 2002.

[55] Y.L. Sandler, "Structure of PWR Primary Corrosion Products," *Corrosion*, Volume 35, Number 5, pp. 205–208, 1979.

[56] L.J. Parrot and D.C. Killoh, "Carbonation in 36 Year Old, In-Situ Concrete," *Cement and Concrete Research*, Volume 19, pp. 649–655, 1989.

[57] P.A. Slegers and P.G. Rouxhet, "Carbonation of the Hydration Products of Tricalcium Silicate." *Cement and Concrete Research*, Volume 6, pp.381–388, 1976.

[58] P.C. Singer and W. Stumm, "Oxygenation of Ferrous Iron: The Rate Determining Step In The Formation of Acid Mine Drainage," Federal Water Pollution Control Administration Research Series Report, National Technical Information Service Report No. PB-189-233, 1968.

[59] P.C. Singer and W. Stumm, "Acid Mine Drainage: The Rate Determining Step," *Science*, Volume 167, pp. 1121–1123, 1970.

[60] V. Jain, L. Yang and K. Chiang, "Chemical Speciation, Using Thermodynamic Modeling, during a Representative Loss-of-Coolant Accident Event," Center for Nuclear Waste Regulatory Analyses Report CWNRA 2004-07, Revision 1, Center for Nuclear Waste Regulatory Analyses, San Antonio, TX, 2004.

[61] V.H. Troutner, "Uniform Aqueous Corrosion of Aluminum—Effects of Various Ions," HW-50133, Hanford Atomic Products Operation, Richland, WA, 1957. This report can be accessed at http://www2.hanford.gov/ddrs/search/RecordDetails.cfm?AKey=D8506811.

[62] C. Groot, "Phosphoric Acid Corrosion Inhibition in Present Reactors," HW-50219, Hanford Atomic Products Operation, Richland, WA, 1957. This report can be accessed at http://www2.hanford.gov/ddrs/common/findpage.cfm?AKey=D8370590.

[63] L.L. Shreir, R.A. Jarman, and G.T. Burstein. *Corrosion*, Volume 1, "Metal/Environment Reactions," Butterworth-Heinemann, Oxford, UK, 1994.

[64] M.G. Fontana, *Corrosion Engineering*, McGraw-Hill, New York, NY, 1986.

[65] D.D. Macdonald, "Characterization of Chemical and Corrosion Effects Potentially Occurring Inside a PWR Confinement Following a LOCA," Appendix E to P.A. Torres, "Peer Review of GSI-191 Chemical Effects Testing Program," NUREG-1861, U.S. Nuclear Regulatory Commission, Washington, DC, December 2006 (ADAMS Accession No. ML063630498).

[66] T.S. Andreychek, M.D. Coury, A.E. Lane, K.F. McNamee, R.B. Sisk, D. Mitchell, P.V. Pyle, W.A. Byers, K.J. Barber, D.P. Crane, M.E. Nissley, D.J. Fink, G. Wissinger, H. Dergel, B.G. Lockamon, and P.A. Sherburne, "Evaluation of Long-Term Cooling Considering Particulate, Fibrous and Chemical Debris in the Recirculating Fluid," WCAP-16793-NP, Revision 0, Westinghouse Electric Company, LLC, Pittsburgh, PA, May 2007 (ADAMS Accession No. ML071580139).

[67] Safety evaluation by the NRR on Topical Report WCAP-16406-P, Revision 1, "Evaluation of Downstream Sump Debris Effects in Support of GSI-191," PWR Owners Group Project No. 694, December 21, 2007 (ADAMS Accession No. ML073520295).

[68] NRC Information Notice 93-66, "Switchover to Hot Leg Injection Following a Loss of Coolant Accident in Pressurized Water Reactors," U.S. Nuclear Regulatory Commission, Washington, DC, August 16, 1993.

[69] W.J. Krotiuk, D. Helton, and C. Boyd, "Post-LOCA PWR Core Inlet Blockage Assessment," NRC Office of Nuclear Regulatory Research, Safety Margins Branch Analysis Report, U.S. Nuclear Regulatory Commission, Washington, DC, February 2007 (ADAMS Accession No. ML070860521).

[70] *U.S. Code of Federal Regulations*, "Standards for Combustible Gas Control Systems in Light-Water-Cooled Power Reactors," Section 50.44, Title 10.

APPENDIX A

Curriculum Vitae for PIRT Panelist

John Anthony Apps

EDUCATION

Ph.D., 1970 (Geology) Harvard University, Cambridge, Massachusetts.

A.M., 1965 (Geology) Harvard University, Cambridge, Massachusetts.

B.Sc. (First Class Honours) 1961 (Mining Geology), A.R.S.M., Imperial College of Science and Technology, London University, England.

EXPERIENCE (Approximately last 20 Years)

SENIOR SCIENTIST EMERITUS, Geochemistry Department, 01/26/04 - present. Earth Science Division, Lawrence Berkeley National Laboratory, University of California, Berkeley, California.

Investigator to evaluate the geochemistry of the unsaturated zone, fate of cementitious materials, fate and transport of long-lived actinides and the temperature effects on radionuclide adsorption at the planned DOE high-level nuclear waste repository at Yucca Mountain,

Investigator with K. Pruess and Xianfu Xu to evaluate CO_2 sequestration in deep saline aquifers (DOE-OBES)

Co-editor with Chin-Fu Tsang of a book on underground injection science and technology for the U.S. EPA.

STAFF SCIENTIST, TERM, Geochemistry Department, Earth Science Division, Lawrence Berkeley National Laboratory, University of California, Berkeley, California. 10/01/01 – 09/30/03 (Returned to full-time employment status)

Investigator to evaluate the geochemistry of the unsaturated zone at the planned DOE high-level nuclear waste repository at Yucca Mountain, Nevada. Investigator to model the geochemistry of the Drift Scale Test at Yucca Mountain (DOE-YMP)

Investigator with K. Pruess and Xianfu Xu to evaluate CO_2 sequestration in deep saline aquifers (DOE-OBES).

Co-convenor with Chin-Fu Tsang of an International Conference on Underground Injection Science and Technology (UIST) for the U.S. EPA

SENIOR SCIENTIST EMERITUS, Geochemistry Department, 06/98 – 09/30/01. Environmental Remediation Technology Department, 02/94 – 06/98. Earth Science Division, Lawrence Berkeley National Laboratory, University of California, Berkeley, California.

Investigator to evaluate the geochemistry of the unsaturated zone at the planned DOE high level nuclear waste repository at Yucca Mountain, Nevada. Investigator to model the geochemistry of the Drift Scale Test at Yucca Mountain (DOE-YMP)

Investigator with K. Pruess and Xianfu Xu to evaluate CO_2 sequestration in deep saline aquifers (DOE-OBES)

Co-principal investigator (with George Moridis), Containment of Contaminants Through Physical Barriers from Viscous Liquids Emplaced Under Controlled Viscosity Conditions (DOE-EM-50);

Co-principal investigator (with Chin-Fu Tsang), Investigation of the Geochemistry of Hazardous Waste Disposal by Deep Underground Injection, (EPA-ODW); and the Calico Hills Project (DOE-YMP).

STAFF SENIOR SCIENTIST, Geochemistry Group, Earth Sciences Division, Lawrence Berkeley Laboratory, University of California, Berkeley, California, 10/82 - 10/31/93. (Elected early retirement).

Principal or co-principal investigator on the following projects:

Coordination with Russian scientists with respect to radioactive remediation at Cheliabinsk, Russia, (with Chin-Fu Tsang), (DOE-EM and DOE-OBES);

Investigation of the geochemistry of hazardous waste disposal by deep underground injection, (EPA-ODW);

Modeling the leaching of fly ash, (EPRI); Oxygen consumption in compressed air energy storage reservoirs, (EPRI)

Evaluation of ground and surface water contamination from mining operations in California, (The State of California);

Investigation of near field rock-water interactions and radionuclide transport, (NRC-Research);

Abiogenic methane formation, (DOE-OBES);

Geochemical aspects of the seismic stability of fractures, (DOE)

Appointed VISITING PROFESSOR in the Department of Material Science and Mineral Engineering, University of California, Berkeley, 8/85

RECENT HONORS

Outstanding Performance Award, Lawrence Berkeley National Laboratory, 1996

Article I. INTERNAL COMMITTEES

Lawrence Berkeley Laboratory Staff Committee, 1982-1985.

Earth Science Division Professional Staff Committee, 1978-1987, 1990-present.

Lawrence Berkeley Laboratory Library Committee, 1984-1987.

Lawrence Berkeley Laboratory Technical Advisory Committee for the Environmental Restoration Program, 1994-present.

PUBLICATIONS

About 50 peer reviewed papers, numerous Government Agency, Laboratory and Company reports, co-editor of two books, and two patents.

EXTRA-MURAL ACTIVITIES (Approximately last 20 Years)

National Research Council, Board of Mineral Resources. Member, Committee on Accessory Elements. Chairman, Panel on Non Bauxitic Aluminum Ores, 1978-1979.

National Research Council, Committee on Radioactive Waste Management. Member, Panel on Savannah River Plant Wastes, 1979-1982.

American Chemical Society, Petroleum Research Fund Advisory Board, 1978-1981. Chairman, Geosciences Committee, 1980-1981. Member, Policy Committee, 1980-1981.

Department of Energy. Member, OBES Earth Science Advisory Committee, (Sewing Circle), 1979-1982.

Department of Energy, Office of Nuclear Waste Isolation. Member, Engineering Review Group, 1983-1986.

Department of Energy, Office of Technology Development, Facilitator, Soil Washing Workshop, Las Vegas, NV, August 28-29, 1990.

Nuclear Regulatory Commission, Advisory Committee on Nuclear Waste Isolation, 1989-1993

Department of Energy, Fernald Environmental Management Project. Uranium Soils Integrated Demonstration Advisory Group: Waste Treatment and Disposal, 1991-1994.

American Institute of Hydrology, Organizing Committee for the 4th Joint USA/CIS Conference on Hydrology, San Francisco, Nov 8-11, 1999

Consulted for Rockwell-Hanford Operations (BWIP), the Office of Nuclear Waste Isolation, (US DOE), US Nuclear Regulatory Commission, US Environmental Protection Agency, US Department of Justice, E.I. Du Pont de Nemours Inc., ALCOA, URS Corp., Dames and Moore, Woodward-Clyde and others.

Wu Chen

Sr. Specialist
Dow Chemical, U. S. A.
B-1603, Freeport, TX 77541
(979) 238-9943
(979) 238-0651 (FAX)
wuchen@dow.com

EDUCATION

Ph.D.	Chemical Engineering	1986	University of Houston
M.S.	Chemical Engineering	1984	University of Houston
B.S.	Chemical Engineering	1979	National Taiwan University

EXPERTISE
- Solids/liquid Separation (filtration, sedimentation, centrifugation and business development)
- Solid handling (storage, transport)
- Particle formation, Characterization, Coagulation and flocculation
- Slurry handling
- Process development, Project management, Bioprocessing, Bioseparation
- Certified 6-Sigma black belt
- People Skill, Leadership training, Marriage counseling

AFFILIATIONS

* **Dow Chemical, U. S. A. (1988-PRESENT)**

Engineering & Process Science, Solid Processing, Freeport, TX (1993 - present)
Project Leader, Research Leader, Sr. Specialist
- Established the solid handling lab. in Freeport (lab. set up, development, equipment procurement, and project management).
- Lead in all aspects of solid handling (fundamentals, tool/guideline development, laboratory evaluation, pilot test, process design, capital project, and plant trouble shooting)

Industrial Biotechnology, San Diego, CA (2001 - 2003)
- Led the implementation of α-Amylases manufacturing process - process development, project management, and plant start up
- Imported/exported equipment and material, for process implementation in Mexico City.

Resins Research, Freeport, TX (1991 - 1993)
Project Leader
- Conducted process research for epoxy products.
- Served as a subject matter expert for global solid separation.

Epoxy Manufacturing, Freeport, TX (1988 - 1991)
Sr. Process Engineer
- Conduct process development and plant trouble shooting.
- Implemented centrifuge & other process technologies into production plants.

* **University of Houston (1986-1988)**

Research Associate
- Managed solid liquid separation lab.
- Led research in mechanism of flow through compressible, porous beds in solid-liquid separation operations-- sponsored by the U. S. Dept. of Energy.

PROFESSIONAL ASSOCIATIONS INVOLVEMENT

American Filtration & Separation Society
- Awarded Well Shoemaker Award - for significant contributions in leadership and service (2005).
- Awarded Frank Tiller Award - for significant technical achievement (2002)
- Served as the society Chairman (2003)
- Served on Board of Directors (1998-2004)
- Served as the Education Committee Chairman (2004-present)
- Served as National conference chair (1996, 2000)

American Institute of Chemical Engineers
- Served as the course leader/instructor for the solid/liquid separation course (1995-present)

U.S. National Renewable Energy Laboratory (NREL)
- Served as a consultant to evaluate the cellulosic ethanol process (2007)

U.S. Nuclear Regulatory Commission
- Served as a consultant for technology review (2005-2006)

Life Strategy
- Served as in instruct on personality/people skill (1999-present)
- Served as a counselor on leadership (2000- present)
- Served as an Instructor/counselor on marriage (1998-present)
  * course offerings include: Houston, Shanghai, WuHan and Beijing.

SELECTED PUBLICATIONS

1. **Chen, W.** and W. P. Li, "Expression", Perry's Chem. Eng. Handbook, 8th Edition, to be published (2008)

2. Tiller, F. M., W. P. Li and **W. Chen**, "Solid-liquid Separation," Chapter 14, Chemical Engineering Handbook, Chapman & Hall, to be published (2007).

3. **Chen, W.** "Analyses of Compressible Suspensions for an Effective Filtration and Deliquoring", Drying Technology J. 24, 1251-1256 (2006).

4. Kraipech, W, **W. Chen** and T. Dyakowski,"The Performance of the Empirical Models on Industrial Hydrocyclone Design", International J. of Mineral Processing, 80, 100-115 (2006).

5. **Chen, W.**, F. Parma and W. Schabel,"Testing Methods for Belt Filter Press Biosludge Dewatering", Filtration, Vol. 5, No. 1 (2005).

6. **Chen, W.**,"Selecting Membrane Filtration System", Cover Story, Chemical Eng. Progress, Dec. (2004).

7. **Chen, W.**,"The Use of Hydrocyclone Models in Practical Design", 9th World Filtration Congress, April 19-22, New Orleans, LA (2004).

8. **Chen, W.**, F. J. Parma, A. Patkar, A. Elkin and S. Sen, "How to Select a Membrane Filtration System," AFS Annual Conference, June 18-20, Reno, NV (2003).

9. Kraipech, W. and **W. Chen**, "Modeling the Fish-Hook Effect of the Flow within Hydrocyclones," AFS Annual Conference, May 1-4, Tampa, Florida (2001).

10. **Chen, W.**, N. Zydek, F. Parma, "Evaluation of Hydrocyclone Models for Practical Applications," Chem. Eng. J., Vol 80, 295-303 (2000).

11. **Chen, W.** and A. Probst, "Scale-up of Gravity Clarifiers and Thickeners from a Dynamic Simulator," AFS Annual Conference, March 14-17, Myrtle Beach, South Carolina (2000).

12. Keller, K. and **W. Chen**, "Solid-liquid Separation: Addressing the Crisis," Chem. Processing, Vol. 63, No.1, Jan. (2000).

13. Tiller, F. M. and **W. Chen**, "How to Use Filter Aids," Chem. Processing, June (1999).

14. **Chen, W.**, "Filter Media Selection from an End User's Point of View," Keynote Speech, Filtration 98, International Conference of INDA, Dec. 3, Atlantic City, New Jersey (1998).

15. **Chen, W.**, "Solid-liquid Separation via Filtration," Chem. Eng., Cover Story, Feb. (1997).

16. **Chen, W.**, "Sedimentation," Chapter 14, Handbook of Powder Science and Technology, 2nd Ed., Chapman & Hall (1997).

17. **Chen, W.**, "The Measurement of Cake Compressibility and Its Application in Chemical Industry," the 7th World Filtration Congress, May 20-23, Budapest, Hungary (1996).

Calvin H. Delegard

Professional Activities

Twenty-five years (1970; 1972-1976; 1979-1987; 1992-present) in applied/process chemistry and nuclear materials safeguards at the Hanford Site.

- Plutonium process and waste chemistry (speciation, separation, purification, processing)
- PUREX and Plutonium Finishing Plant process chemistry
- Environmental chemistry of radionuclides (Co, Sr, Tc, Cs, U, Np, Pu, Am)
- Technical liaison with the Institute of Physical Chemistry, Russian Academy of Sciences, on chemistry of the actinides and technetium in alkaline media, on K Basin sludge treatment/disposal, chromium phase dissolution, and uranium metal corrosion
- Hanford tank waste chemistry (bulk and radionuclide components)
  - Tank farm evaporator chemistry (concentration and crystallization)
  - Organic decomposition reactions in tank waste (e.g., gas generation in 101-SY)
  - Tank waste treatment chemistry (ozonation, electrolysis, calcination/dissolution, leaching/sludge washing)
- Material corrosion; Zr and other metals in nitric acid; copper and mild steel in basalt groundwater
- Chemical characterization, processing, and U metal corrosion of K Fuel Storage Basins sludge
- Domestic (234-5) and design of int'l. (Hanford, RFES, SRS) nuclear materials safeguards
- Development and qualification of plant calorimeters for international safeguards
- Development of prompt gamma analysis for Pu materials characterization
- Plutonium materials stabilization for storage.

Eight years (1976-79; 1987-92), nuclear materials safeguards at the International Atomic Energy Agency, Vienna, Austria.

- International nuclear material safeguards inspection execution
- Subsidiary arrangements, design information, and facility attachments preparation
- Destructive and non-destructive nuclear materials verification methods development.

Technical Publications

Contractor and National Laboratory Documents
- Atlantic Richfield Hanford Company
- Rockwell Hanford Operations
- Westinghouse Hanford Company
- B&W Hanford, Duke ESH, B&W Protec, Numatec Hanford
- Los Alamos National Laboratory
- Brookhaven National Laboratory
- Pacific Northwest National Laboratory

Symposium Proceedings
- Institute of Nuclear Science, Boris Kidrič, Vinča, Yugoslavia
- Institute of Nuclear Materials Management

- American Nuclear Society
- American Society Metallurgical Engineers
- Waste Management
- Materials Research Society
- NATO Advanced Research Workshop
- American Chemical Society
- International Atomic Energy Agency
- Minerals, Metals & Materials Society
- Russian Conference on Radiochemistry
- European Safeguards Research and Development Association

Scientific Journals
- Radiochimica Acta
- Radiochemistry
- Nuclear and Chemical Waste Management
- Journal of Alloys and Compounds
- Journal of Nuclear Material

Book Chapter – in <u>Advances in Plutonium Chemistry 1967-2000</u>

Selected Recent Reports and Articles

- Budantseva, N., A. Bessonov, I. Tananaev, A. Fedosseev, and C. H. Delegard, 1998, "Behavior of Plutonium(V) in Alkaline Media," *Journal of Alloys and Compounds* 271-273:813-816.
- Budantseva, N. A., I. G. Tananaev, A. M. Fedosseev, and C. Delegard, 1998, "Capture of Pu(V), Np(V) and Pu(VI) from Alkaline Solutions by Hydroxides of Pu(IV), Th(IV) and La(III)," *Journal of Alloys and Compounds* 271-273:231-235.
- Peretroukhine, V. F., and C. H. Delegard, 1998, "Some Comparisons of Plutonium-Bearing Radwaste Management in the USA and Russia," <u>The Environmental Challenges of Nuclear Disarmament</u>, NATO Advanced Research Workshop, Cracow, Poland.
- Barney, G. S. and C. H. Delegard, 1999, "Chemical Species of Plutonium in Hanford Site Radioactive Tank Wastes," in <u>Actinide Speciation in High Ionic Strength Media</u>, D. T. Reed, S. B. Clark, and L. Rao (editors), Kluwer Academic/Plenum Publishers, New York.
- Bredt, P. R., C. D. Carlson, C. H. Delegard, K. H. Pool, A. J. Schmidt, D. B. Bechtold, J. Bourges, D. A. Dodd, T. A. Flament, N. N. Krot, and A. B. Yusov, 2000, "Studies of Chemical Processing of K Basin Sludge," <u>Waste Management 2000</u>, Tucson, AZ.
- Delegard, C. H., S. A. Bryan, A. J. Schmidt, P. R. Bredt, C. M. King, R. L. Sell, L. L. Burger, and K. L. Silvers, 2000, "Gas Generation from K East Basin Sludges – Series I Testing," PNNL-13320, Pacific Northwest National Laboratory, Richland, WA. (http://www.pnl.gov/main/publications/external/technical_reports/PNNL-13320.pdf).
- Bryan, S. A., C. H. Delegard, A. J. Schmidt, R. L. Sell, K. L. Silvers, S. R. Gano, and B. M. Thornton, 2001, "Gas Generation from K East Basin Sludges – Series II Testing," PNNL-13446, Pacific Northwest National Laboratory, Richland, WA. (http://www.pnl.gov/main/publications/external/technical_reports/PNNL-13446Rev1.pdf).
- Delegard, C., 2001, "K Fuel Basin Sludge – Characterization, Reactions, Retrieval, and Storage," presentation to Washington Group International, Warrington, UK, October 9, 2001, PNWD-SA-5513, Pacific Northwest Division, Battelle, Richland, WA.
- Fedoseev, A. M., V. P. Shilov, N. A. Budantseva, A. B. Yusov, I. A. Charushnikova, and C. H. Delegard, 2002, "Selective Recovery of Chromium from Precipitates Containing d Elements and Actinides: I-III. Effect of O_2, H_2O_2, $S_2O_8^{2-}$," *Radiochemistry* 44(4):347-365 (three articles).

- Clark, S. B. and C. Delegard, 2002, "Plutonium in Concentrated Solutions," Chapter 7 in <u>Advances in Plutonium Chemistry 1967-2000</u>, D. C. Hoffmann, Editor, American Nuclear Society, La Grange Park, IL (http://www.ans.org/store/vi-300029).
- Delegard C. H., S. I. Sinkov, B. K. McNamara, S. A. Jones, G. S. Barney, A. J. Schmidt, and R. L. Sell, 2002, "Critical Mass Laboratory Solutions – Precipitation, Calcination, and Moisture Uptake Investigations," PNNL-13934, Pacific Northwest National Laboratory, Richland, WA (http://www.pnl.gov/main/publications/external/technical_reports/PNNL-13934.pdf).
- Boak, J., E. A. Conrad, C. H. Delegard, A. M. Murray, G. D. Roberson, and T. J. Venetz, 2003, "Recommendations on Stabilization of Plutonium Material Shipped to Hanford from Rocky Flats," LA-UR-03-3789, Los Alamos National Laboratory, Los Alamos, NM (http://lib-www.lanl.gov/cgi-bin/getfile?00937076.pdf).
- Schmidt, A. J., C. H. Delegard, S. A. Bryan, M. R. Elmore, R. L. Sell, K. L. Silvers, S. R. Gano, and B. M. Thornton, 2003, "Gas Generation from K East Basin Sludges and Irradiated Metallic Uranium Fuel Particles – Series III Testing," PNNL-14346, Pacific Northwest National Laboratory, Richland, WA (http://www.pnl.gov/main/publications/external/technical_reports/PNNL-14346.pdf).
- Fazzari D. M., S. A. Jones, and C. H. Delegard, 2003, "Application of Prompt Gamma-Ray Analysis to Identify Electrorefining Salt-Bearing Plutonium Oxide at the Plutonium Finishing Plant," PNNL-14409, Pacific Northwest National Laboratory, Richland, WA (http://www.pnl.gov/main/publications/external/technical_reports/PNNL-14409.pdf).
- Delegard C. H., A. J. Schmidt, R. L. Sell, S. I. Sinkov, S. A. Bryan, S. R. Gano, and B. M. Thornton, 2004, "Final Report - Gas Generation Testing of Uranium Metal in Simulated K Basin Sludge and in Grouted Sludge Waste Forms," PNNL-14811, Pacific Northwest National Laboratory, Richland, WA (http://www.pnl.gov/main/publications/external/technical_reports/PNNL-14811.pdf)
- Yusov, A. B., A. M. Fedosseev, and C. H. Delegard, 2004, "Hydrolysis of Np(IV) and Pu(IV) and Their Complexation by Aqueous Si(OH)$_4$," *Radiochimica Acta* 92(12):869-881.
- Shilov, V. P., A. M. Fedoseev, A. B. Yusov, and C. H. Delegard, 2004, "Behavior of Np(VII, VI, V) in Silicate Solutions," *Radiochemistry* 46(6):574-577.

Education

BS, Chemistry, with distinction, Phi Beta Kappa, Washington State University, 1970; Graduate course work at the University of Colorado (1970-72) and Washington State University (1973-present).

calvin.delegard@pnl.gov, (509) 376-0548
Pacific Northwest National Laboratory
PO Box 999, MS P7-25
Richland, WA 99352 USA

Robert Litman

28 Hutchinson Drive
Hampton, NH 03842
(603) 926-4863
drbob20@comcast.net

EDUCATION
1. Brooklyn College - B.S., Chemistry (1971)
2. City University of New York - PhD, Analytical Chemistry (1975)

PROFESSIONAL EXPERIENCE

2002 to now -	Independent Consultant
1998 to 2002 -	Principal Chemist (Engineering), Seabrook Station
1996 to 1998 -	Chemistry Manager, Seabrook Station
1985 to 1996 -	Chemistry Supervisor, Seabrook Station
1981 to 1985 -	Senior Chemistry Training Instructor, Seabrook Station
1975 to 1981 -	Assistant Professor of Chemistry, University of Lowell
1971 to 1975 -	Graduate Fellow 'A' City University of New York at Brooklyn College

additionally...

2002	Presented Technology Transfer Award by EPRI for leadership in development of the Primary Water Chemistry Guidelines
1997	Peer Evaluator at Diablo Canyon Nuclear Power Plant for INPO
1996 to present -	1. Contract Lecturer and Radiochemistry Consultant for RSCS, Inc. as a Senior Chemist 2. Independent Contractor working on the Multi-Agency Radiochemistry Analytical Protocols Manual (MARLAP) For the EPA (via Sanford Cohen & Assoc.1998-9, The Henry Jackson Foundation 1999 and Environmental Management Support, Inc. 2001- present).
1992 to 1996 -	1. Contract Lecturer for Technical Management Services, Inc. Teaching "Practical Radiochemistry" 2. Independent Contractor for USEPA NAREL Montgomery, AL. Reviewed technical procedures for accuracy and modified them for technical enhancements. Performed independent review of laboratory techniques.
1990 to 1992 -	Participated as a Training Instructor in the Water Pollution Control Facility Operator Training Courses given by the State of NH
1992 -	Received Excellence award from Seabrook Station for maintaining and improving chemistry programs.
1993 -	Coordinated and developed the 1993 Summer program for the New Hampshire State Science Teachers Association.

1994 - Appointed as a Peer Evaluator for Seabrook Station Training programs by INPO.

PROFESSIONAL AFFILIATIONS

1. American Chemical Society (since 1971)
2. Sigma Xi, The Scientific Research Society of NA (since 1971)
3. ASTM Committee D19-Water (since 1996)
4. Standard Methods Special Committee on Development of ^{90}Sr Analysis in Water

RECREATIONAL INTERESTS

1. USYSA Soccer Coach, 'D' License
2. USSF Soccer Referee, Level 8 (1999 NH State Referee of the Year)
3. NH State High School Soccer Official
4. ASA Softball Umpire
5. NH State High School Softball Umpire
6. NH Seacoast Youth Soccer League- Vice President (2001 to present)
7. USYS Referee Training Instructor (2003 - present)

AREAS OF TECHNICAL EXPERTISE WHILE AT SEABROOK STATION

I was Seabrook Station's Principal Chemist during my last five years at the plant. My job responsibilities included corrosion control methods, analysis of corrosion mechanisms, environmental innovations for biocide effectiveness, long term trending of plant chemistry performance parameters, monitoring of trends in plant radiochemical parameters, and radiological effluent surveillance oversight.

During my tenure at Seabrook the chemistry programs received the highest ratings from both the Nuclear Regulatory Commission (NRC) and the Institute of Nuclear Power Operations (INPO). We also received the highest ratings from the NHDES for our NPDES program compliance. Part of the technical responsibility that I was responsible for was the NPDES Permit Renewal Process, evaluation of non-routine discharges and program implementation for new biocides and anti-scalants.

One of the programs I initiated at Seabrook was component inspection. This program helped to assess corrosion mechanisms, biological fouling, and effectiveness of general corrosion control. The inspections provided a chronology so that from one maintenance period to the next an accurate assessment cold be made of the components health. The plant engineering group relies on these inspections to help maintain system efficiencies.

Another important program was the integration of the station's primary to secondary leak response. I worked with computer engineering, operations, Instrument and Control, and chemistry personnel to provide control room operators with a continuous monitor, which provide a gallon per day read out, as well as a rate of change display. I also provided the training on the new system to these groups.

I participated in the site Environmental Review Board (ERB) and was a team member for the successful ISO 14001 Certification Program in 2001. I served as the Chairperson of the Laboratory Quality Control and Audit Committee (LQCAC), which each year evaluates the laboratory that the station uses for 10CFR50/61 and Bioassay programs. This committee was comprised of laboratory clients interested in ensuring that the technical programs met regulatory requirements.

I represented the Station on five technical EPRI committees and have made significant contributions in those areas since 1998:

1. Primary Water Chemistry Guidelines
2. Secondary Water Chemistry Guidelines
3. Primary to Secondary Leak Guidelines
4. Stator Coolant System Guidelines
5. Robust Fuels Working Group 1.

AREAS OF RESEARCH/INTEREST

During the six years I was at the University of Lowell, I successfully supervised nine undergraduate senior projects eight masters theses and one PhD thesis. These projects were centered on environmental analysis and basic analytical/radiochemical techniques. During that time I was responsible for initiating a multidisciplinary environmental studies program at the master's level. This included staff members from chemistry, biology, civil engineering, chemical engineering and earth sciences.

I have been an auditor for several nuclear power plant chemistry programs and contract laboratory programs for their radiochemical processes and procedures over the past 15 years. In that time, I have aided in the development and improvement of these procedures.

The 40-hour instructional course in radiochemistry that I teach has been presented 23 times in the past fifteen years. Attendees at these courses included personnel from state and government as well as contract radiochemistry laboratories and nuclear power facilities. This has helped me to maintain a current status of the radiochemical practices in use, and where the radiochemistry community needs to progress in this area.

Digby D. Macdonald

Distinguished Professor of Materials Science and Engineering
Director, Center for Electrochemical Science and Technology
The Pennsylvania State University
201 Steidle Bldg.
University Park, PA 16802
(814) 863-7772, (814) 863-4718 (fax), ddm2@psu.edu

(a) EDUCATIONAL BACKGROUND
B.Sc. (1965) and M.Sc. (1966) in Chemistry, University of Auckland (New Zealand);
Ph.D. in Chemistry (1969), University of Calgary (Canada).

PROFESSIONAL EXPERIENCE (past 35 years)
Distinguished Professor of Materials Science and Engineering, 6/2003 – present.
Chair, Metals Program, Penn. State Univ., 6/2001 – 6/2003
Director, Center for Electrochemical Sci. & Tech., Penn. State Univ., 7/99 – present.
Vice President, Physical Sciences Division, SRI International, Menlo Park, CA, 1/98 – 7/99
Director, Center for Advanced Materials, Penn. State Univ., 7/91-3/2000
  Professor, Materials Science and Engineering, Penn. State Univ., 7/91 – 6/03.

Deputy Director, Physical Sciences Division, SRI International, Menlo Park, CA, 4/87 - 7/91
Laboratory Director, Mat. Research Lab., SRI International, Menlo Park, CA, 4/87 – 7/91
Laboratory Director, Chemistry Laboratory, SRI International, Menlo Park, CA, 3/84 – 4/87
Director and Professor, Fontana Corrosion Center, Ohio State University, 3/79 – 3/84
Sr. Metallurgist, SRI International, Menlo Park, CA, 3/77 – 3/79.
Sr. Research Associate, University of Calgary, Canada, 3/75 – 3/77.
Lecturer in Chemistry, Victoria University of Wellington, New Zealand, 4/72 – 3/75.
Assist. Research Officer, Whiteshell Nuclear Research Establishment, Atomic Energy of
  Canada Ltd., Pinawa, Manitoba, Canada, 9/69 – 4/72.

CONSULTING ACTIVITIES (Partial list for the last five years))
Integrated Design Technologies
OLI Systems
Electric Power Research Institute
SRI International
Stone & Webster Engineering Co.
Science Applications International Corp.
Framatome
Metallic Power Inc.
Los Alamos National Laboratory

Section 1.02 PATENTS
1. D. D. Macdonald and A. C. Scott, "Pressure Balanced External Reference Electrode Assembly and Method", US Patent 4,273,637 (1981).
2. D. D. Macdonald, "Apparatus for Measuring the pH of a Liquid", US Patent 4,406,766 (1983).

3. S. C. Narang and D. D. Macdonald, Novel Solid Polymer Electrolytes", US Patent 5,061,581 (1991).
4. S. Hettiarachchi, S. C. Narang, and D. D. Macdonald, "Synergistic Corrosion Inhibitors Based on Substituted Pyridinium Compounds", US Patent 5,132,093 (1992).
5. S. Hettiarachchi, S. C. Narang, and D. D. Macdonald, "Reference Electrode Assembly and Process for Constructing", US Patent, 5,238,553 (1993).
6. D. D. Macdonald, et al, "Conducting Polymer for Lithium/Aqueous Syst.", US Prov. Pat. 60/119,360 (1998).
7. D. D. Macdonald, et al, "Polyphosphazenes as Proton Conducting Membranes", US Pat. Appl. 09/590,985 (1999).
8. D. D. Macdonald, et al, "Impedance/Artificial Neural Network Method...", US Prov. Pat. 60/241,871 (1999)
9. D. D. Macdonald, "Electrochemical Conditioning of Wine", US Prov. Pat. 60/295,080 (2001).

RESEARCH INTERESTS
- Passivity and passivity breakdown on metals
- Deterministic prediction of corrosion damage
- Advanced batteries and fuel cells
- Chemistry and electrochemistry of supercritical aqueous systems
- Nuclear power generation
- Electrocatalysis

(a) **PUBLICATIONS** (from approximately 160 published over the past five years and from a total of 650 published since 1970.).

1. Engelhardt, G.R., M. Urquidi-Macdonald, and D. D. Macdonald. "A Simplified Method for Estimating Corrosion Cavity Growth Rates". Corros. Sci., **39**(3), 419-441 (1997).
2. Gao, L. and D. D. Macdonald. "Characterization of Irreversible Processes at the Li/Poly[bis(2,3-di-(2-methoxyethoxy)propoxy)phosphazene] Interface on Charge Cycling". J. Electrochem. Soc., **144**(4), 1174-1179 (1997).
3. Sikora, E. and D. D. Macdonald. "Defining the Passive State". Solid State Ionics, **94**, 141-150 (1997).
4. Engelhardt, G. R. and D. D. Macdonald, "Deterministic Prediction of Pit Depth Distribution", Corrosion, **54**, 469-479 (1998).
5. Macdonald, D. D., E. Sikora, and J. Sikora, "The Kinetics of Growth of the Passive Film on Tungsten in Acidic Phosphate Solutions", Electrochim. Acta, **43**, 2851-61 (1998).
6. Pensado-Rodriguez, O., J. Flores, M. Urquidi-Macdonald, and D. D. Macdonald, "The Electrochemistry of Lithium in Alkaline Aqueous Electrolytes. Part II: Point Defect Model". J. Electrochem. Soc., **146**, 1326-1335 (1999).
7. Engelhardt, G. R., D. D. Macdonald, and P. Millett, "Transport Processes in Steam Generator Crevices. I. General Corrosion Model", Corros. Sci., **41**, 2165-2190 (1999).
8. Macdonald, D. D., "Passivity: The Key to Our Metals-Based Civilization", Pure Appl. Chem., **71**, 951-986 (1999).
9. Quiroga Becerra, H., D. D. Macdonald, and C. Retamosa, "The Corrosion of Carbon Steel in Oil-in-Water Emulsions Under Controlled Hydrodynamic Conditions", Corros. Sci., **42**, 561-575 (2000).
10. Sikora, J., E. Sikora, and D. D. Macdonald, "Electronic Structure of the Passive Film on Tungsten", Electrochim Acta, **45**(12), 1875-1883 (2000).
11. Sikora, E. and D. D. Macdonald, "The Passivity of Iron in the Presence of EDTA. Part I. General Electrochemical Behavior". J. Electrochem. Soc., **147**(11), 4087-4092 (2000).

12. Macdonald, D. D., M. Al-Rafaie and G. R. Engelhardt, "New Rate Laws for the Growth and Reduction of Passive Films", J. Electrochem. Soc., **148**(9), B343 - B347 (2001).
13. Macdonald, D. D., "Probing the Chemical And Electrochemical Properties of SCWO Systems", Electrochim. Acta, **47**, 775-790 (2001).
14. Zhou, X.Y., S. N. Lvov, X. J. Wei, L. G. Benning, and D. D. Macdonald, "Quantitative Evaluation of General Corrosion of Type 304 Stainless Steel in Subcritical and Supercritical Aqueous Solutions via Electrochemical Noise Analysis". Corros. Sci., **44**(2), 841-860 (2002).
15. Engelhardt, G. and D. D. Macdonald, "unification of the Deterministic and Statistical Approaches for Predicting Localized Corrosion Damage. I. Theoretical Foundation", Corros. Sci., 46, 2755 (2004).

(b) PROFESSIONAL ASSOCIATIONS AND HONORS

- Selector of the Kuwait Prize for Applied Sciences, 1985.
- Research Award, College of Engineering, Ohio State University, 1983.
- The 1991 Carl Wagner Memorial Award from The Electrochemical Society.
- The 1992 Willis Rodney Whitney Award from The National Association of Corrosion Engineers.
- Chair, Gordon Research Conference on Corrosion, New Hampshire, 1992.
- W.B. Lewis Memorial Lecture by Atomic Energy of Canada, Ltd., 1993, "in recognition of [his] contributions to the development of nuclear power in the service of mankind".
- Elected Fellow, NACE-International, 1994.
- Member, USAF Scientific Advisory Board, Protocol Rank: DE-4 (Lieutenant General equivalent), 1993-1997
- Elected Fellow, The Electrochemical Society, 1995.
- Elected Fellow, Royal Society of Canada, 1996. ("National Academy" of Canada).
- Wilson Research Award, College of Earth and Minerals Sciences, Pennsylvania State University, 1996.
- Elected Fellow, Royal Society of New Zealand, 1997. ("National Academy" of New Zealand).
- H. H. Uhlig Award, The Electrochemical Society, 2001.
- U. R. Evans Award, The British Corrosion Institute, 2003.
- Appointed Adjunct Professor, Massey University, New Zealand, 2003.
- Appointed Adjunct Professor, University of Nevada at Reno, 2003.
- Elected Fellow, World Innovation Foundation, 2004.

APPENDIX B

Principal Physical Attributes of the LOCA Accident Sequence

Presented at PIRT Panelist Meeting
San Antonio, Texas March 27, 2006

Principal Attributes of the LOCA Accident Sequence

Bruce Letellier

Nuclear Design and Risk Analysis Group
Los Alamos National Laboratory

Chemical Testing Peer Review Panel
Southwest Research Institute
March 27, 2006

Objectives

- **Recap physical attributes of the accident sequence**
  - Thermal Hydraulics relevant to chemistry
- **Supplement mental image of spatial scale and complexity**
- **Provide reference information for remaining discussions**
- **Resolve system-response question relevant to ICET**
- **Stimulate broader discussion of possible chemical effects**

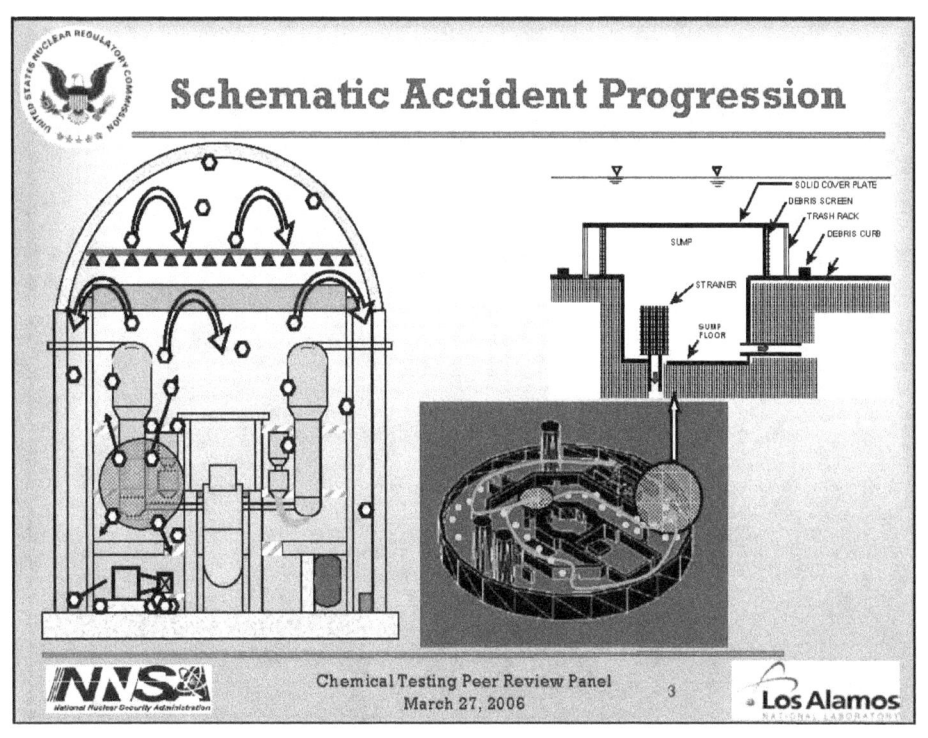

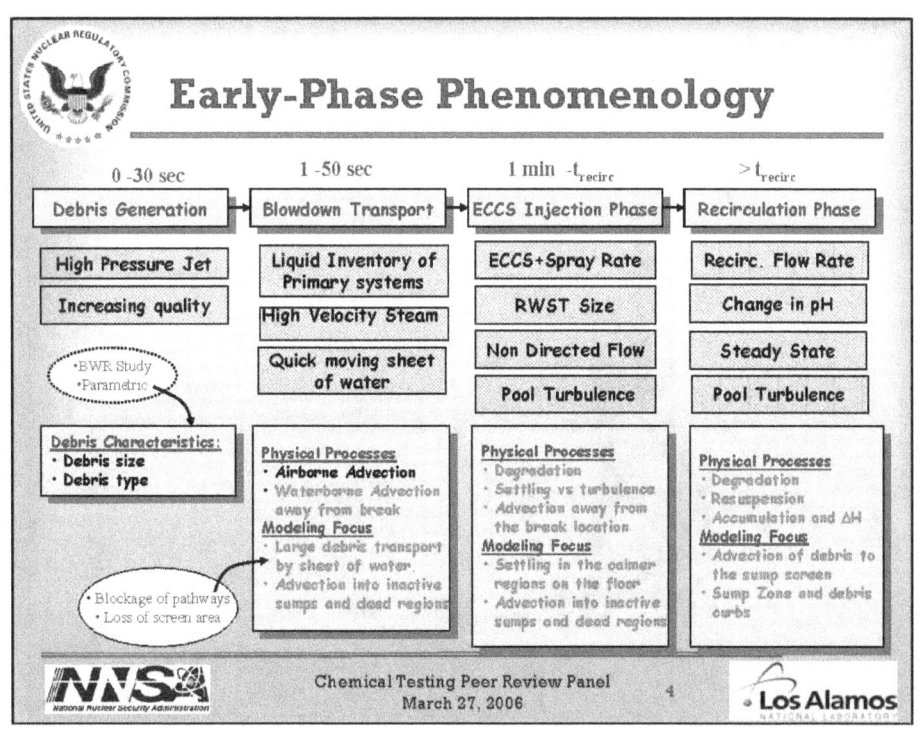

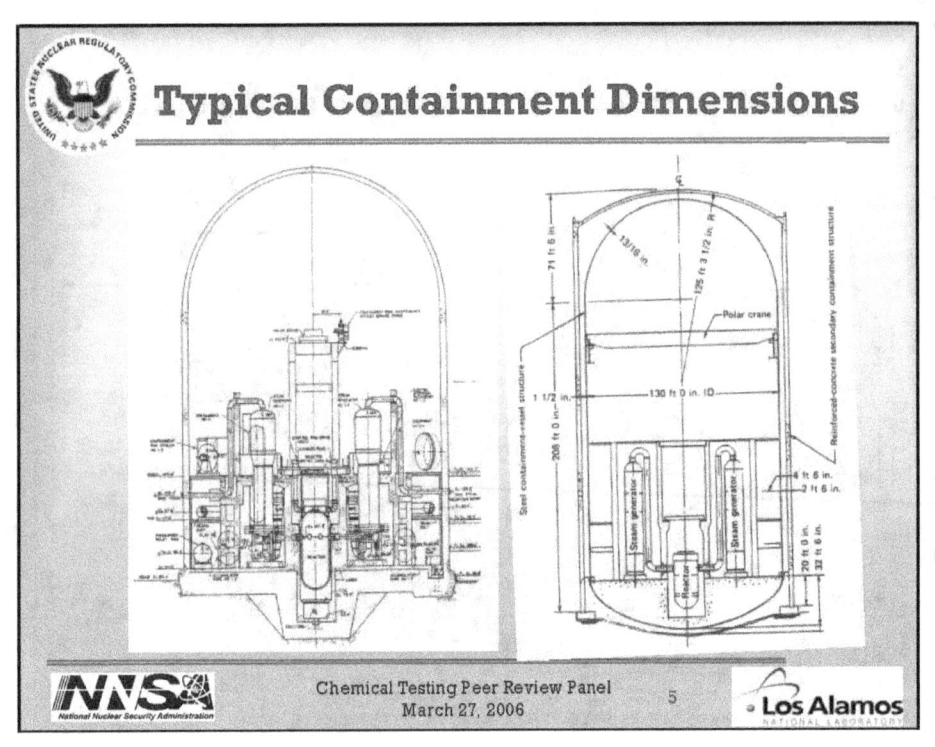

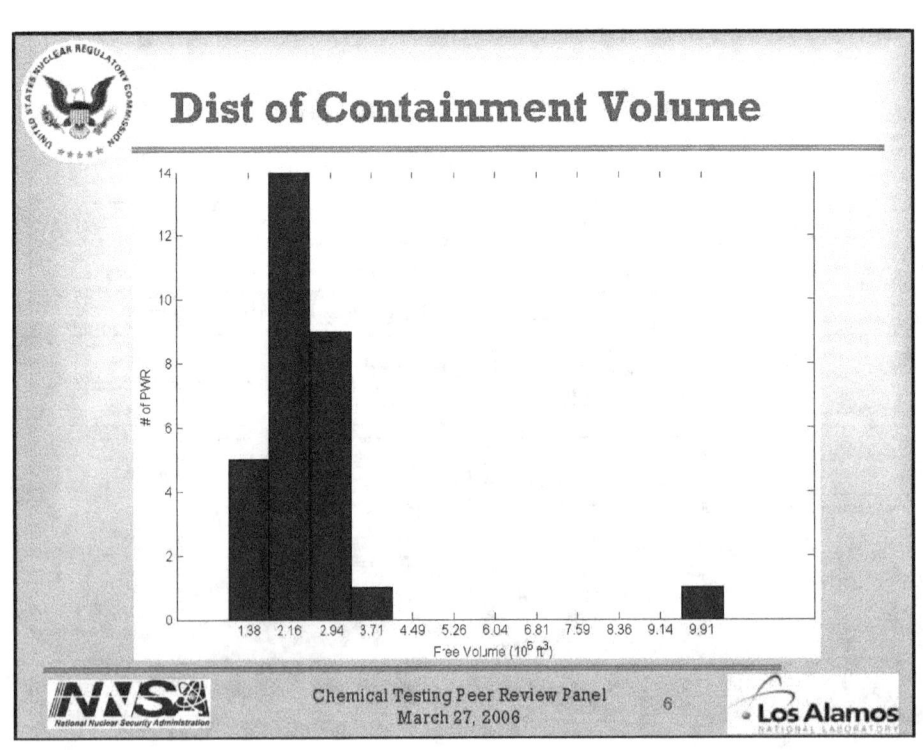

Volunteer Plant Information: Plant Layout

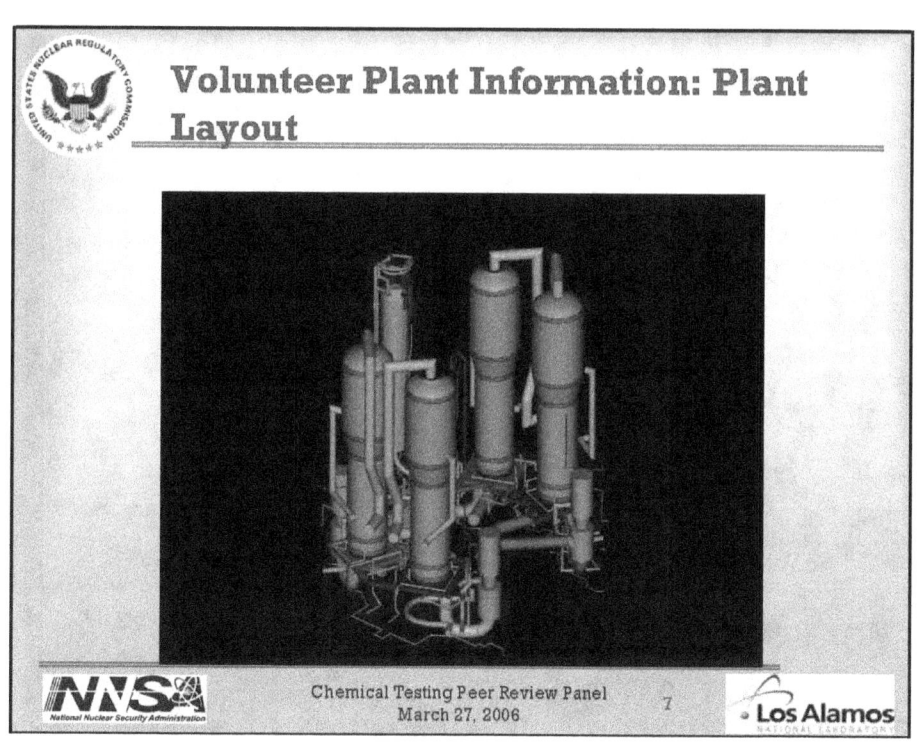

Chemical Testing Peer Review Panel
March 27, 2006

48-foot Diam ZOI at D.C. Cook

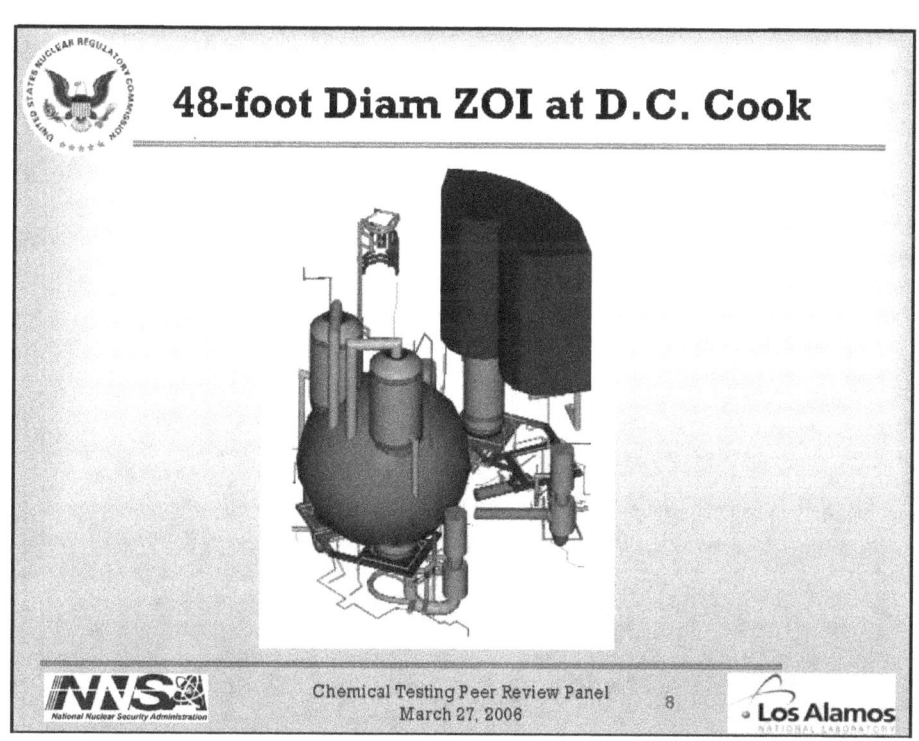

Chemical Testing Peer Review Panel
March 27, 2006

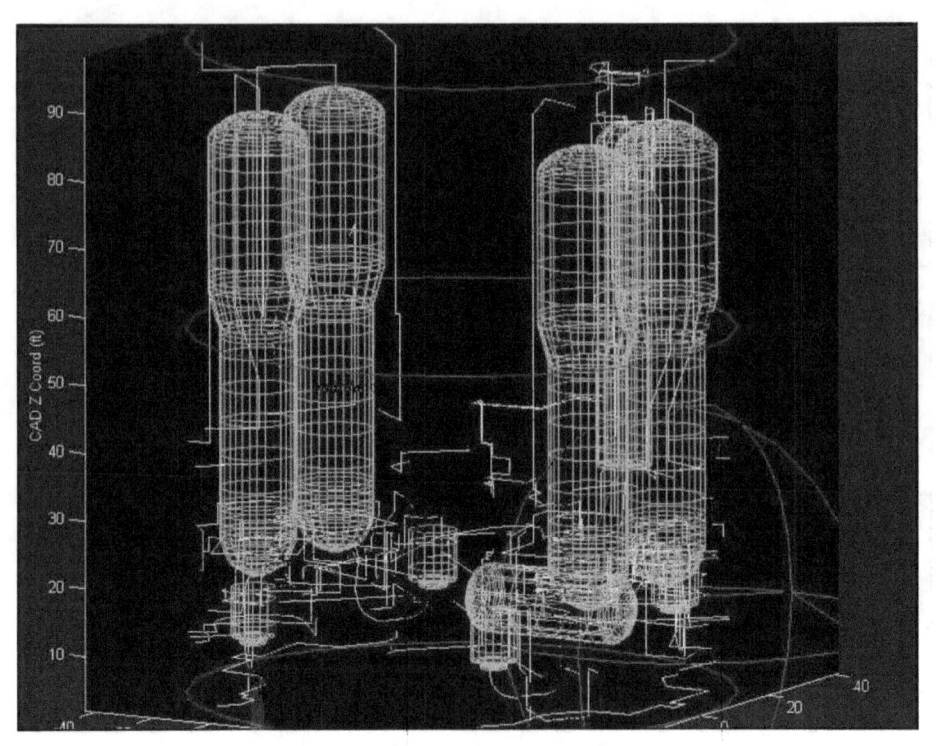

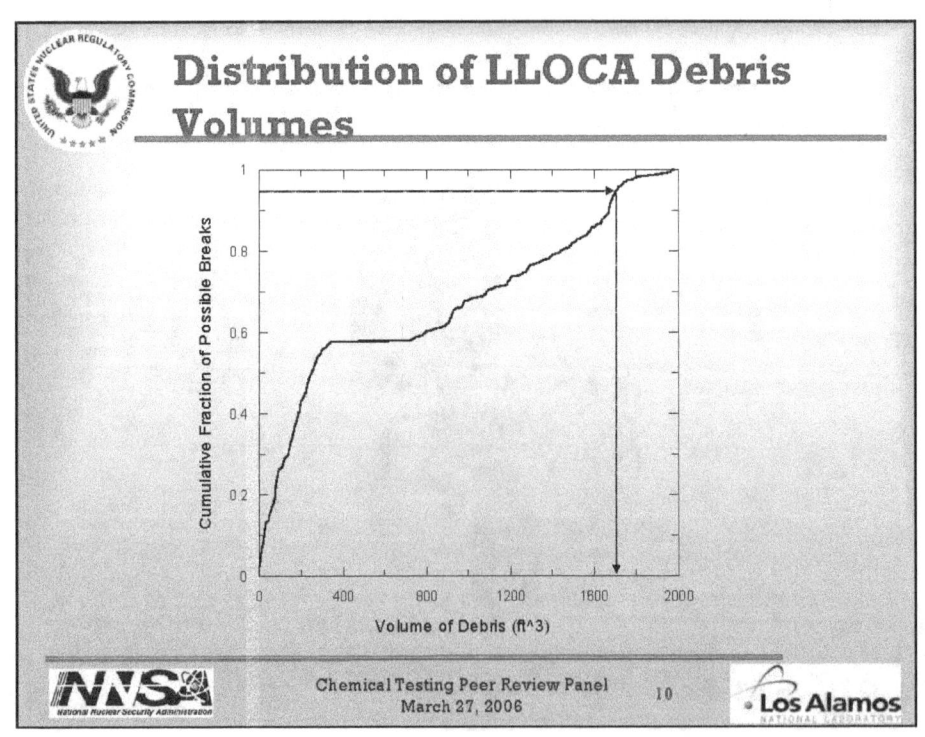

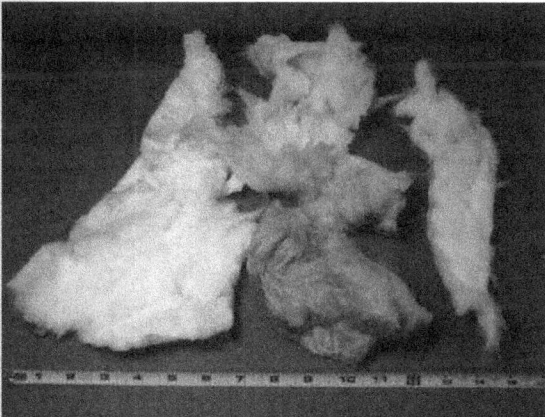

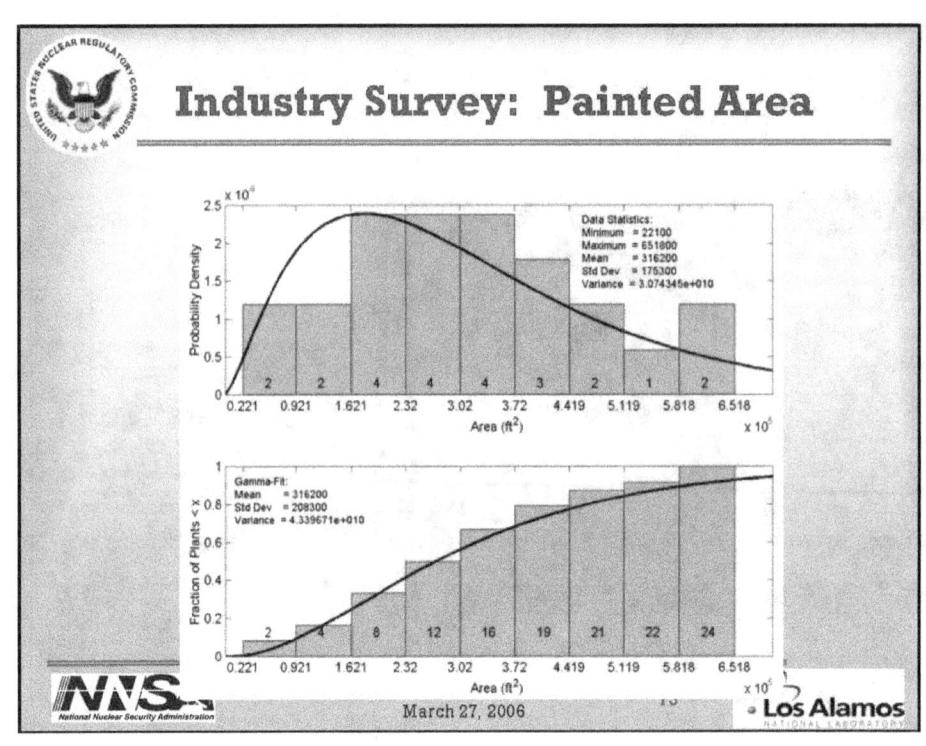
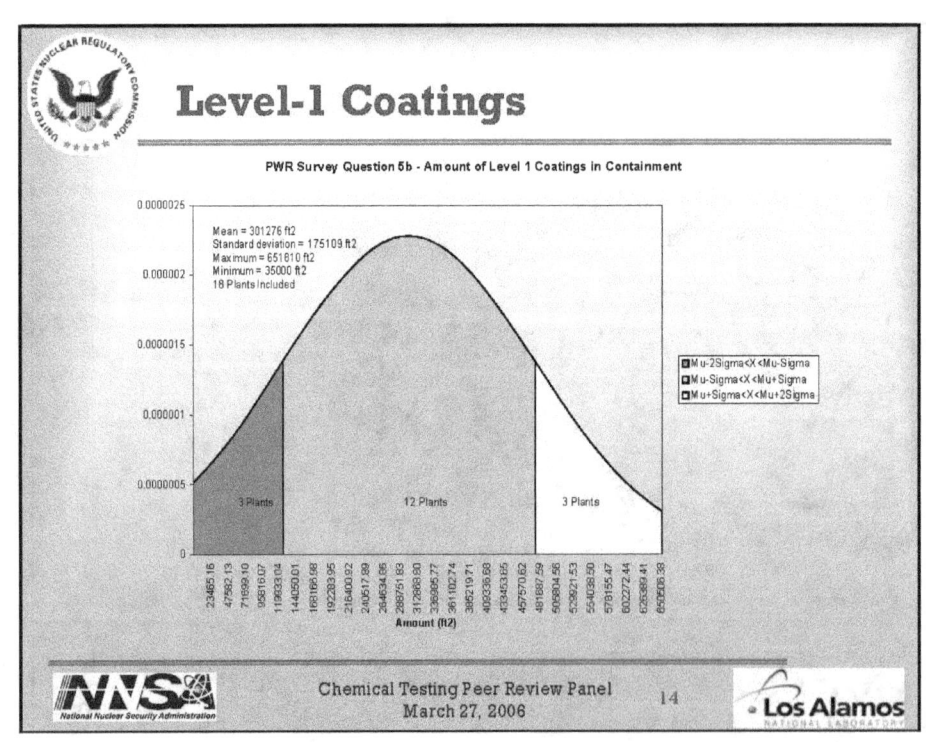

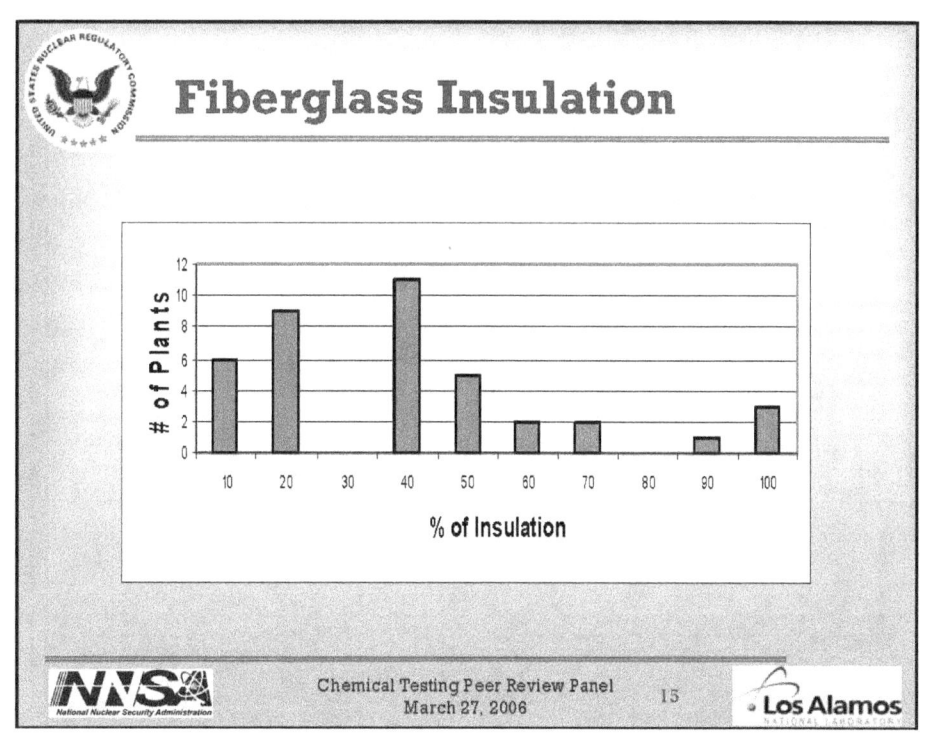

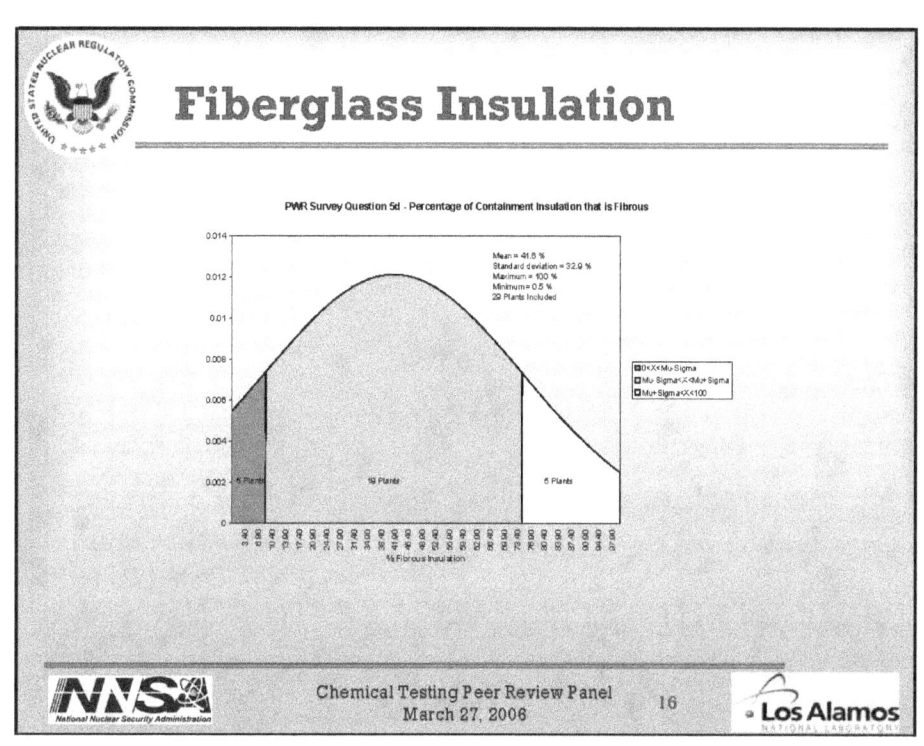

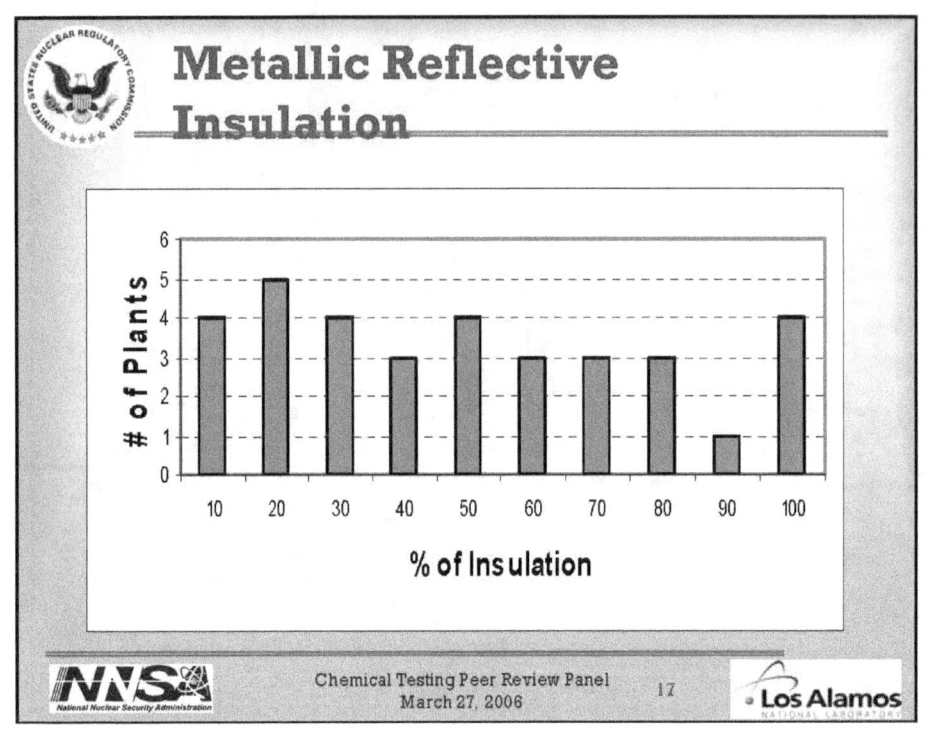
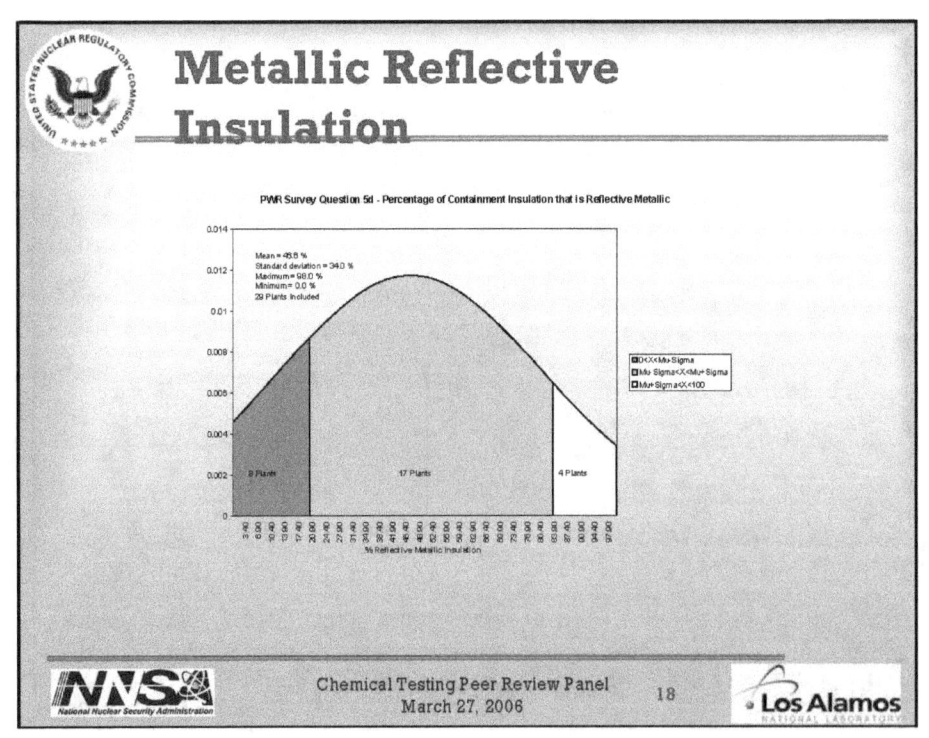

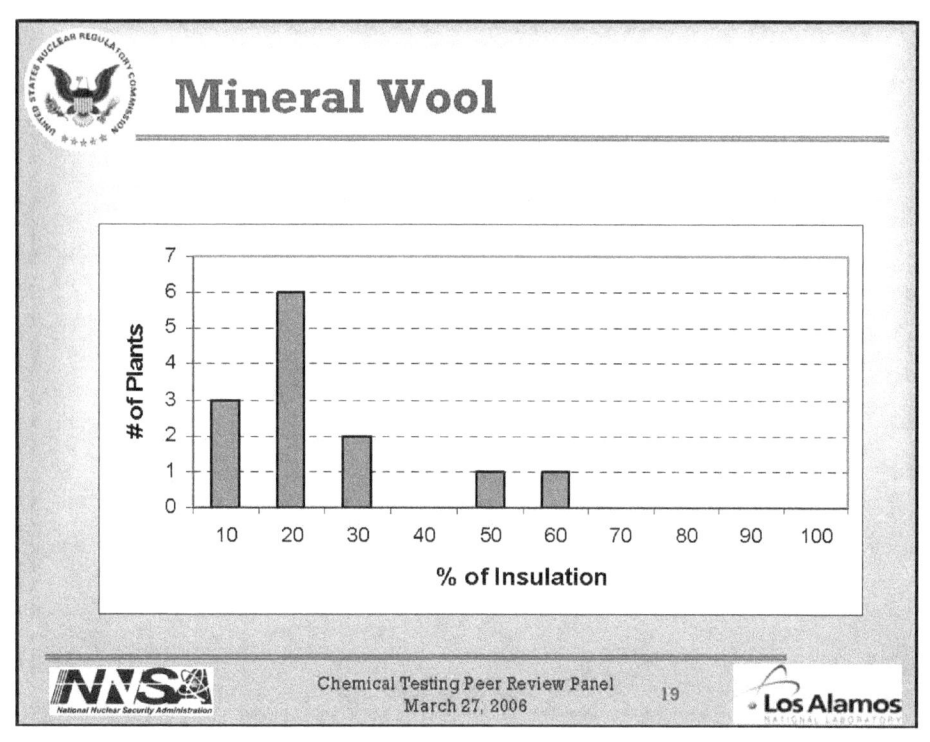
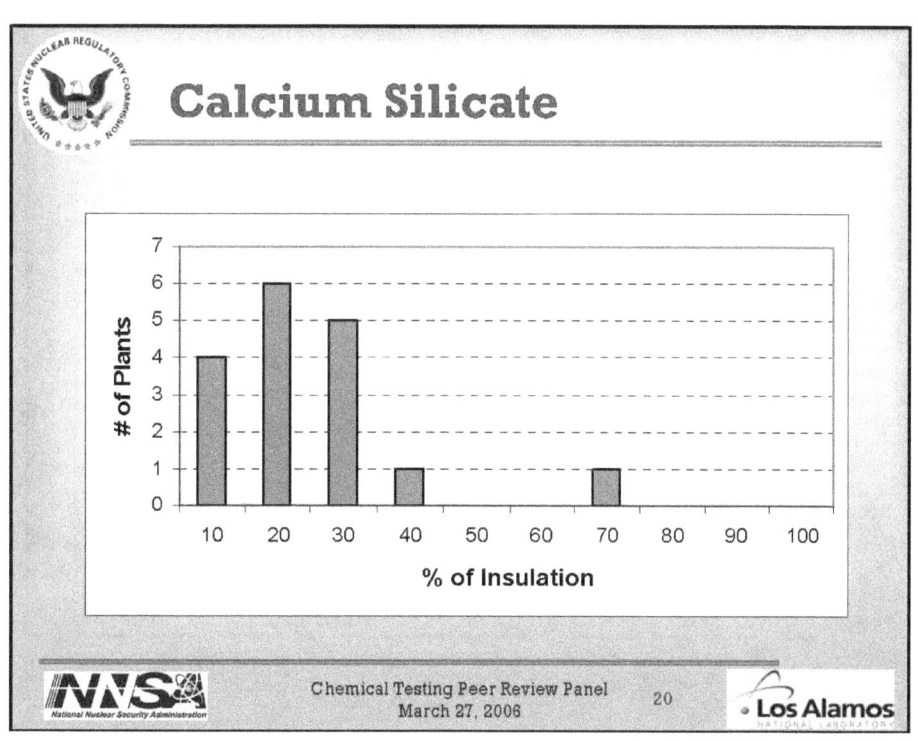

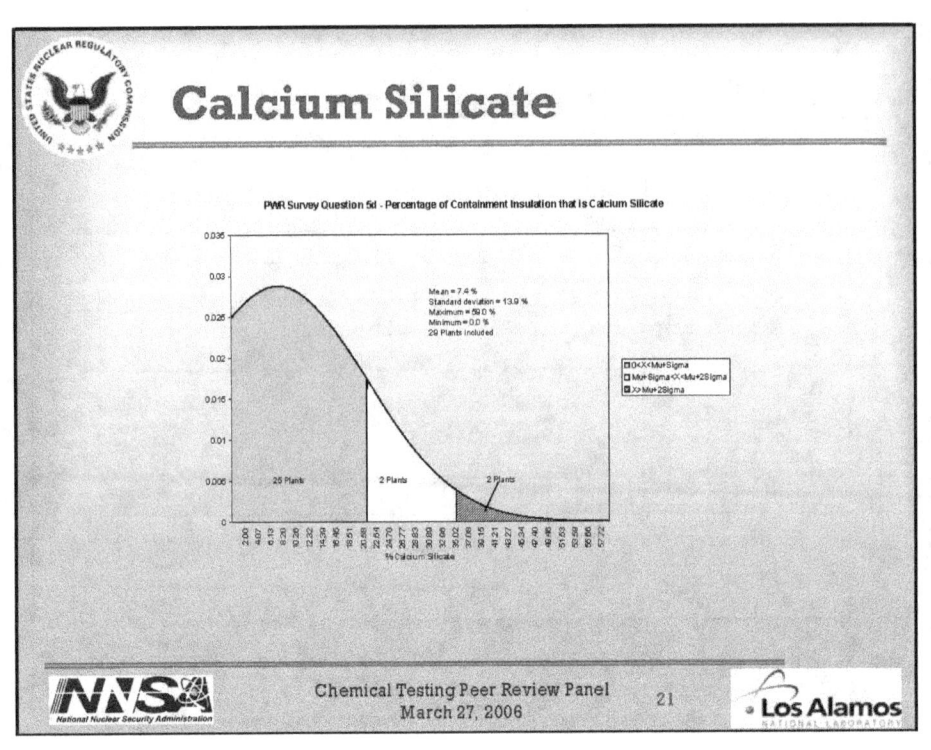

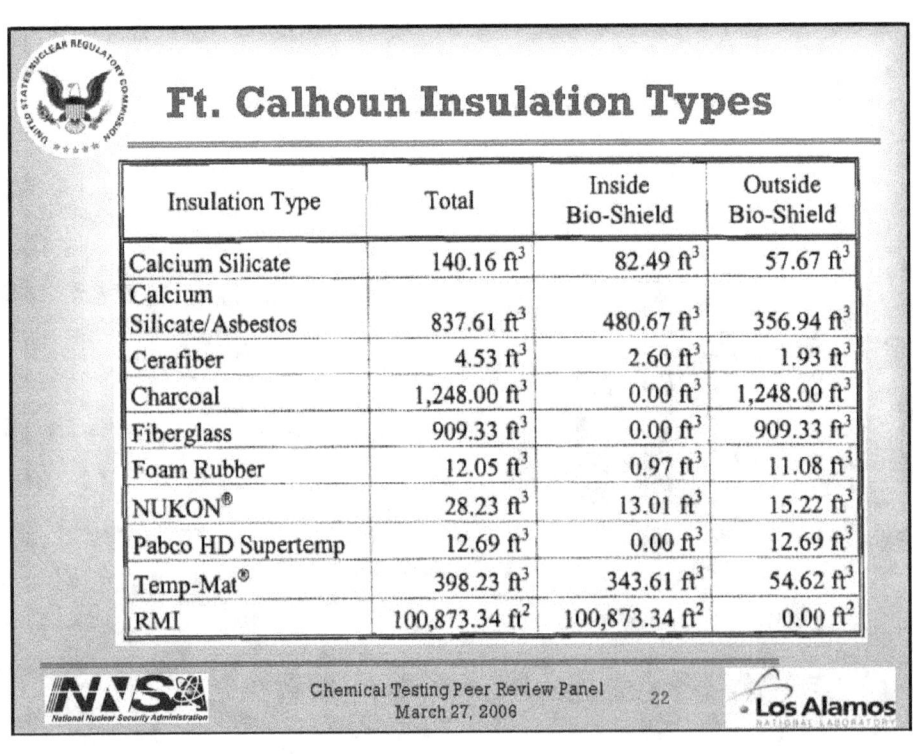

Ft. Calhoun Debris Scenarios

HIGH PARTICULATE SCENARIO

Insulation Type	Quantity Destroyed
Calcium Silicate	0.15 ft^3
Calcium Silicate/Asbestos	93.3 ft^3
Cerafiber	1.72 ft^3
NUKON®	1.33 ft^3
Temp-Mat®	23.30 ft^3
RMI	33,645.68 ft^2

HIGH FIBER SCENARIO

Insulation Type	Quantity Destroyed
Calcium Silicate	4.81 ft^3
Calcium Silicate/Asbestos	49.41 ft^3
Cerafiber	0.88 ft^3
Foam Rubber	0.54 ft^3
NUKON®	2.30 ft^3
Temp-Mat®	162.47 ft^3
RMI	33,645.68 ft^2

Industry-Wide Insulation Types

- NUKON
- NUKON Jacketed
- Fiberglass Blanket
- Fiberglass metallic jacketed
- Fiberglass glass cloth jacketed
- Metallic Reflective
- Stainless Metallic Reflective
- RMI
- Mineral Wool
- Temp Mat
- Calcium Silicate
- Calcium Silicate Jacketed

- Asbestos
- Unibestos
- Kaowool
- Armaflex
- Neoprene
- Vinylcell
- Foamglass
- Transco
- Min "K"
- Transco RMI
- Transco Encapsulated

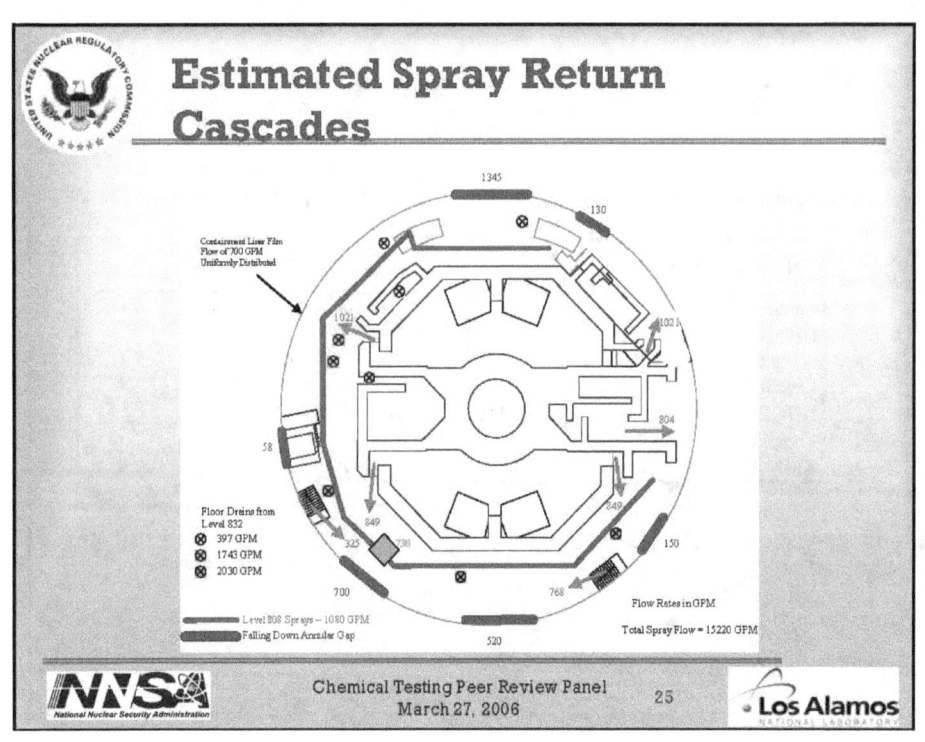

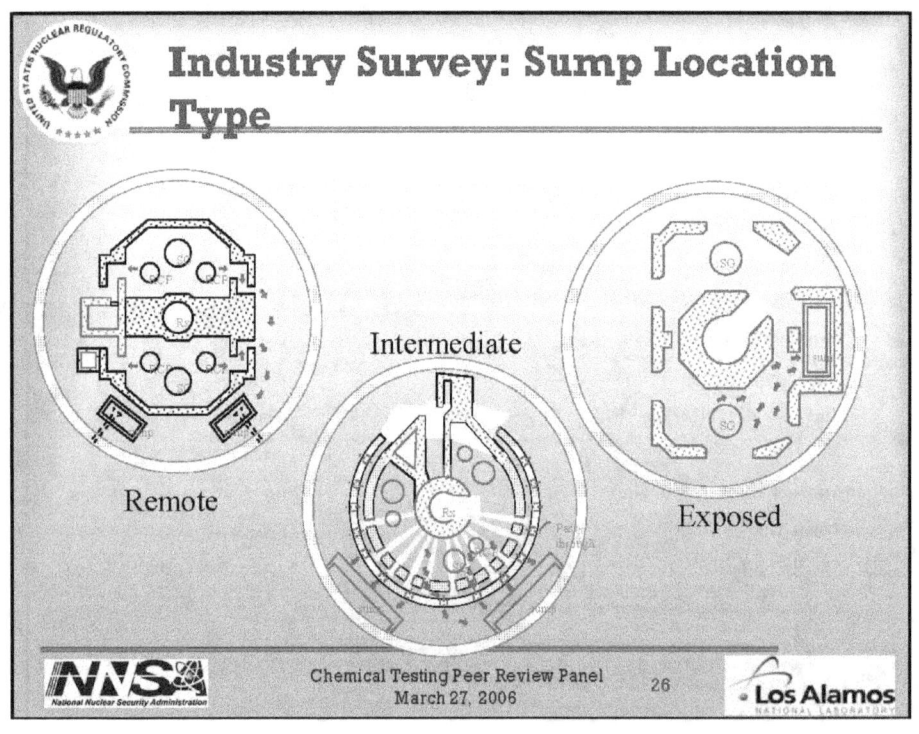

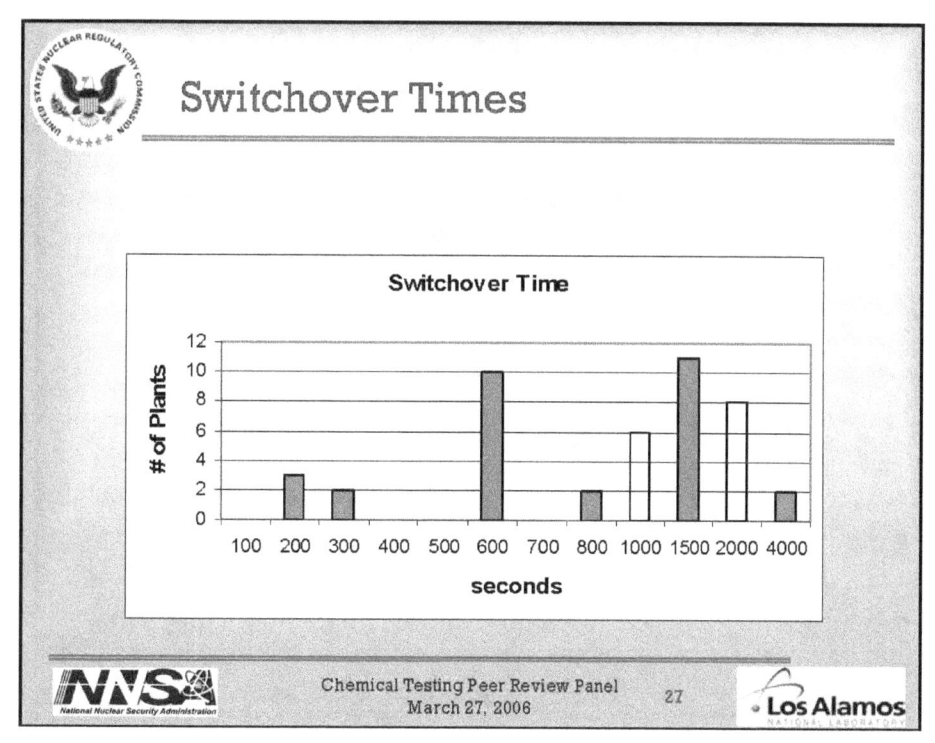

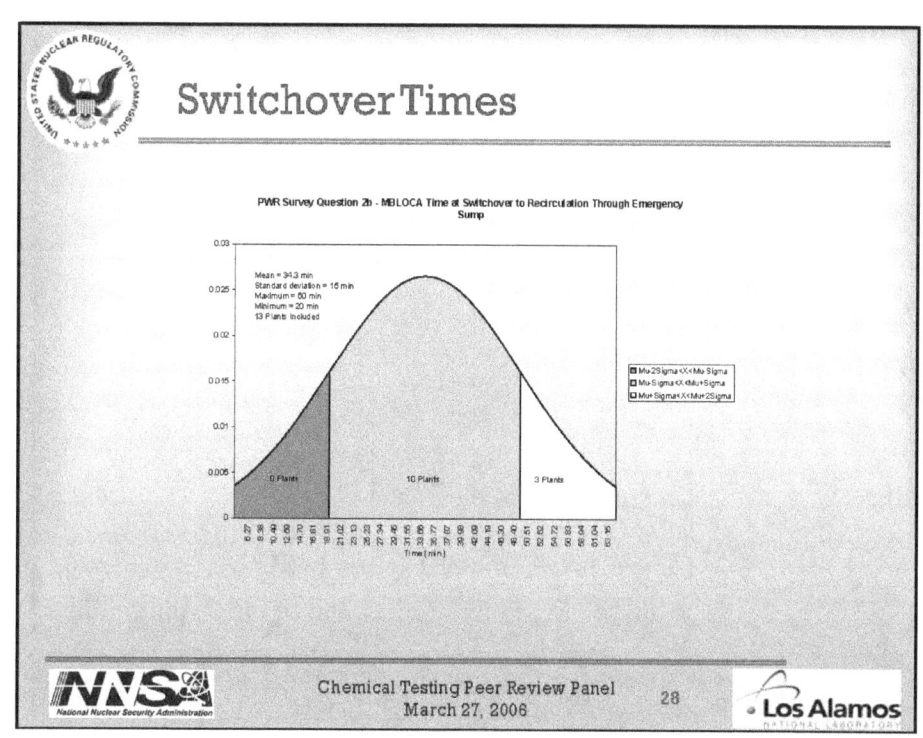

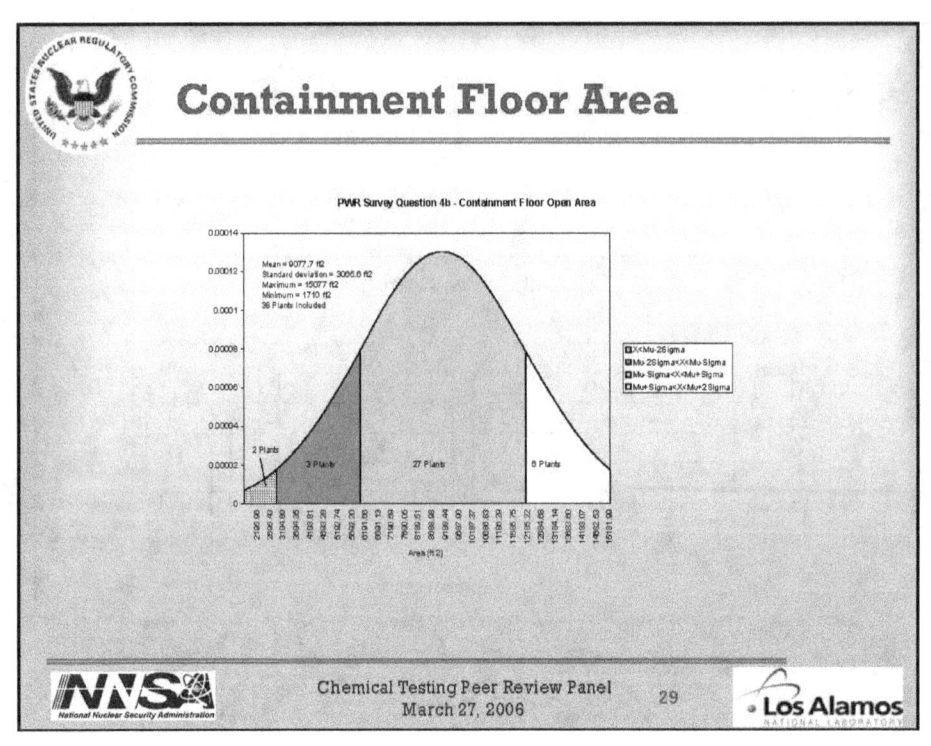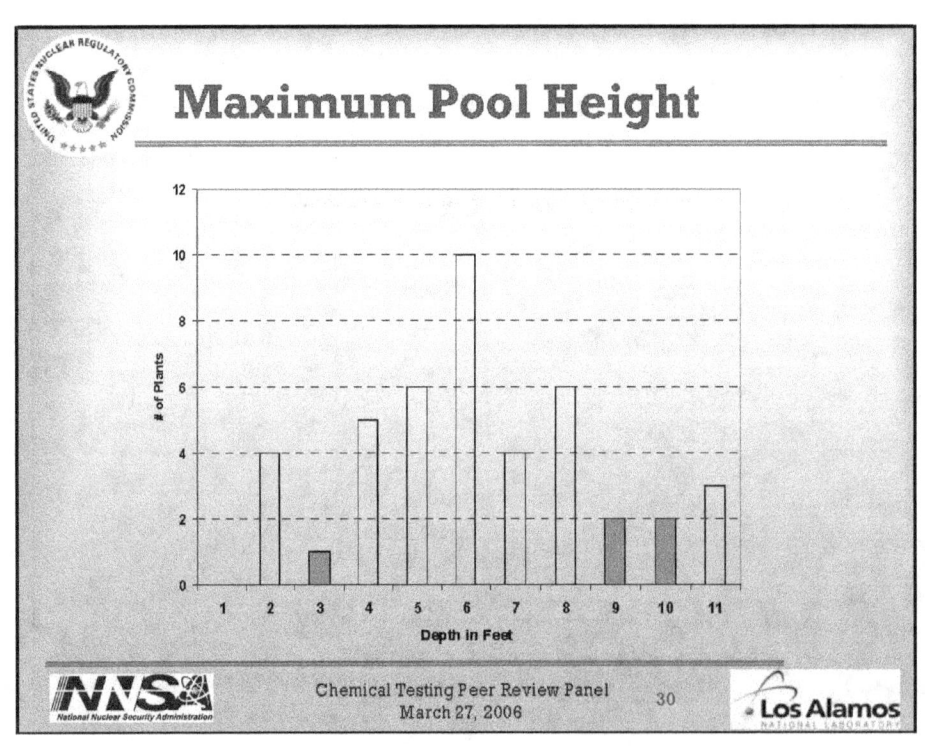

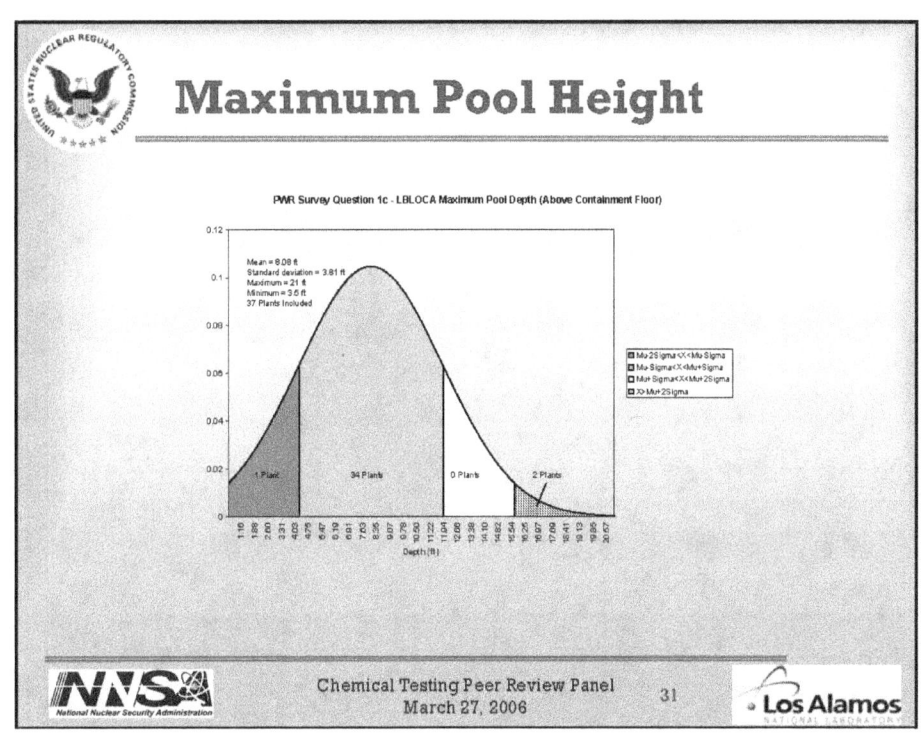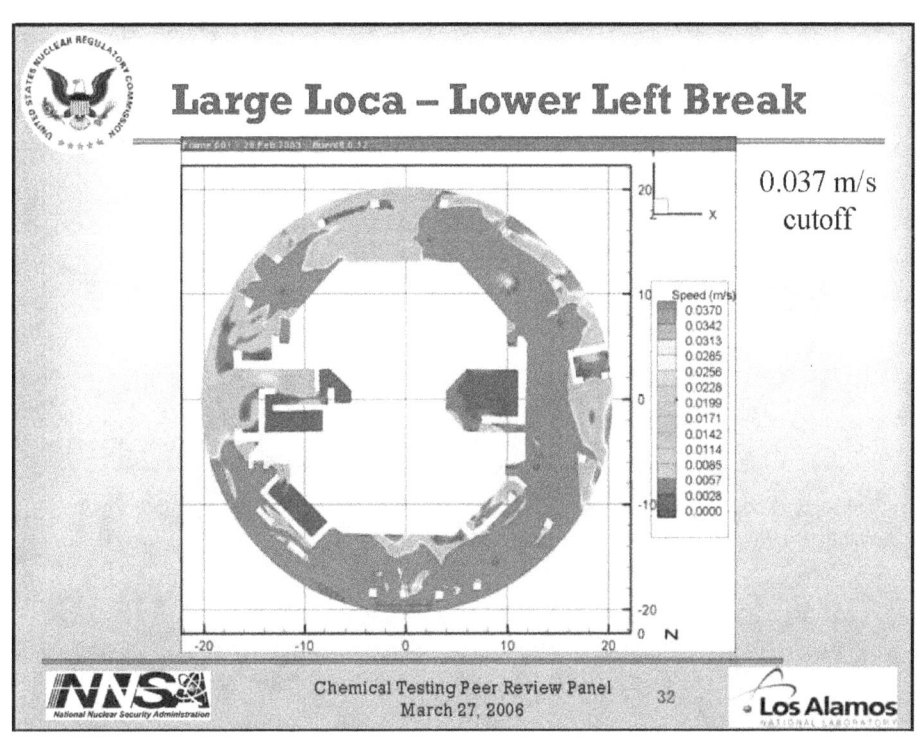

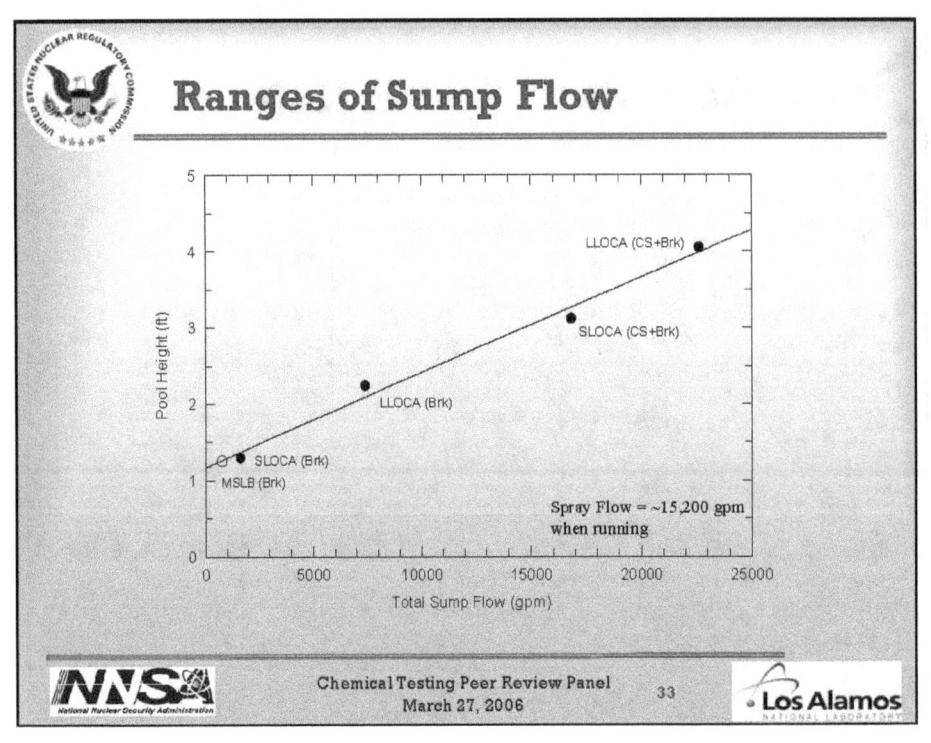

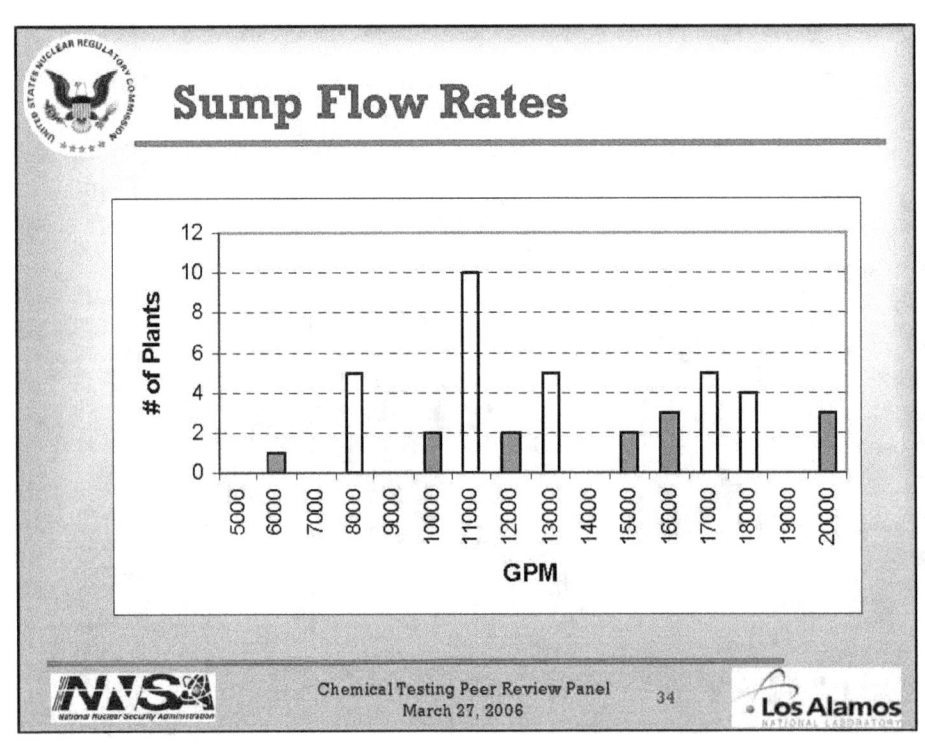

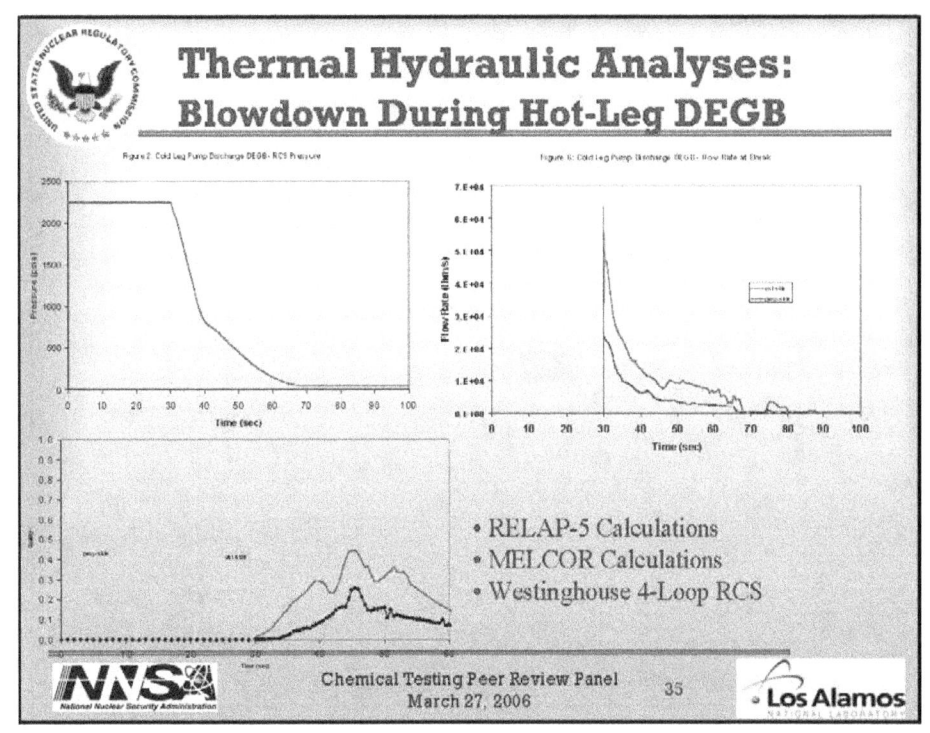

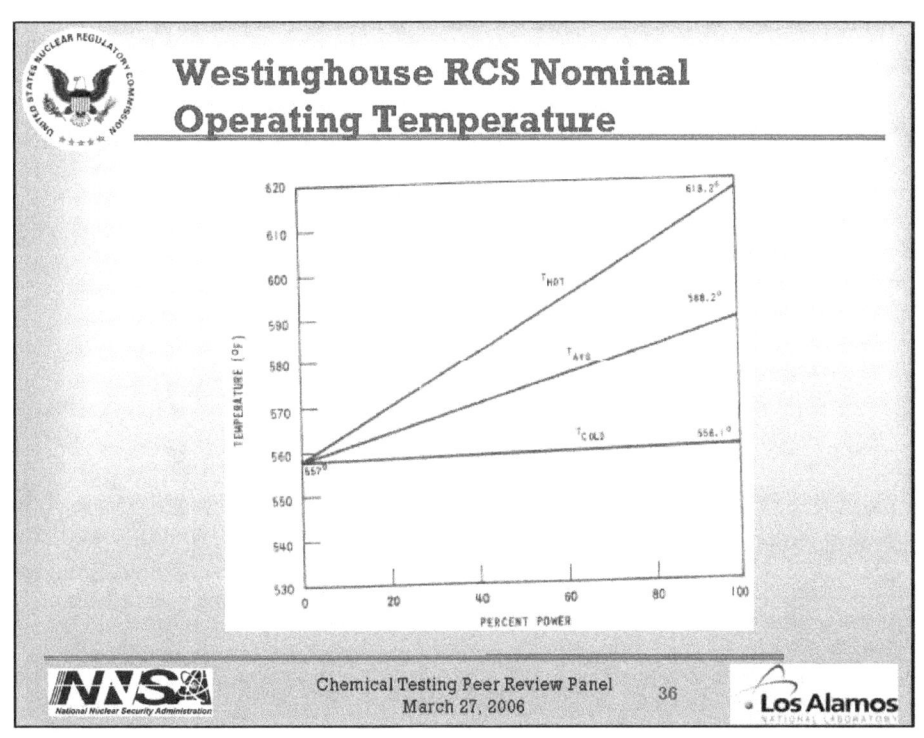

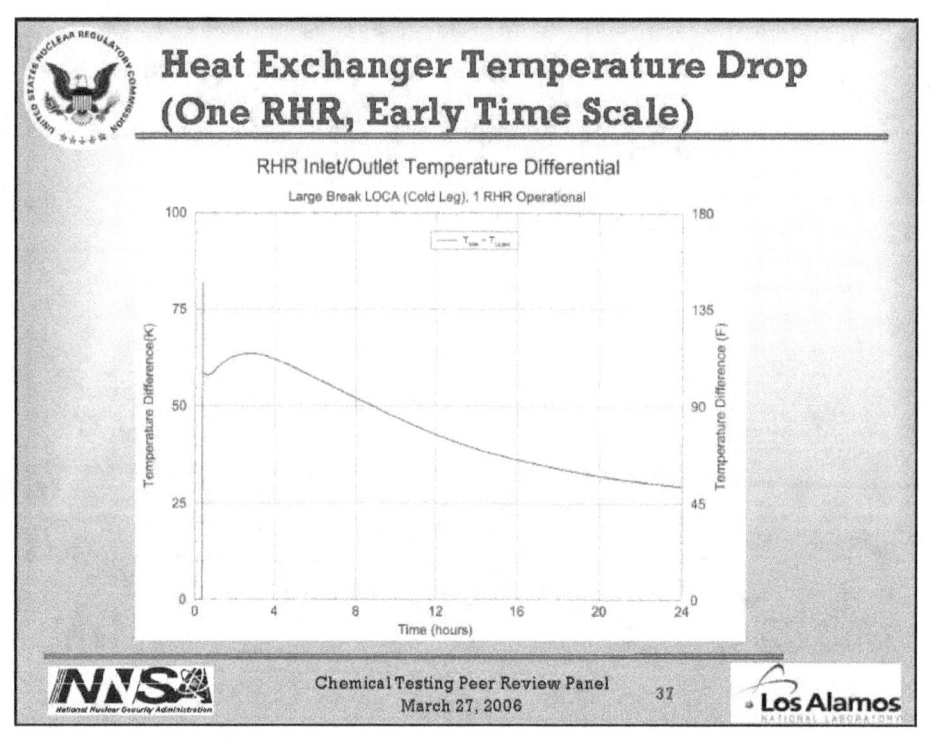
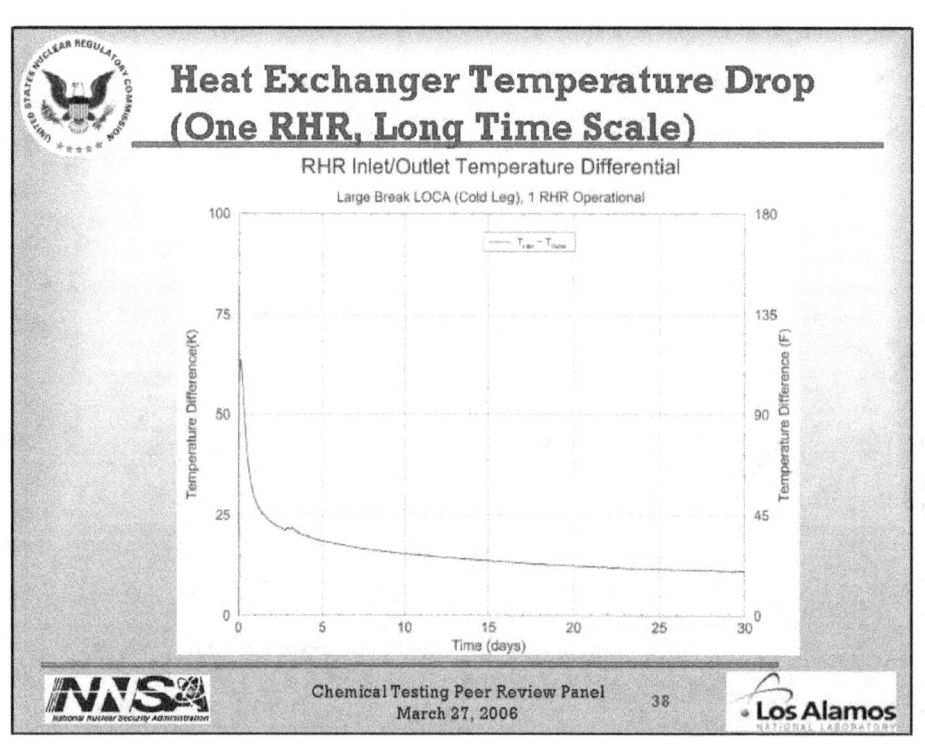

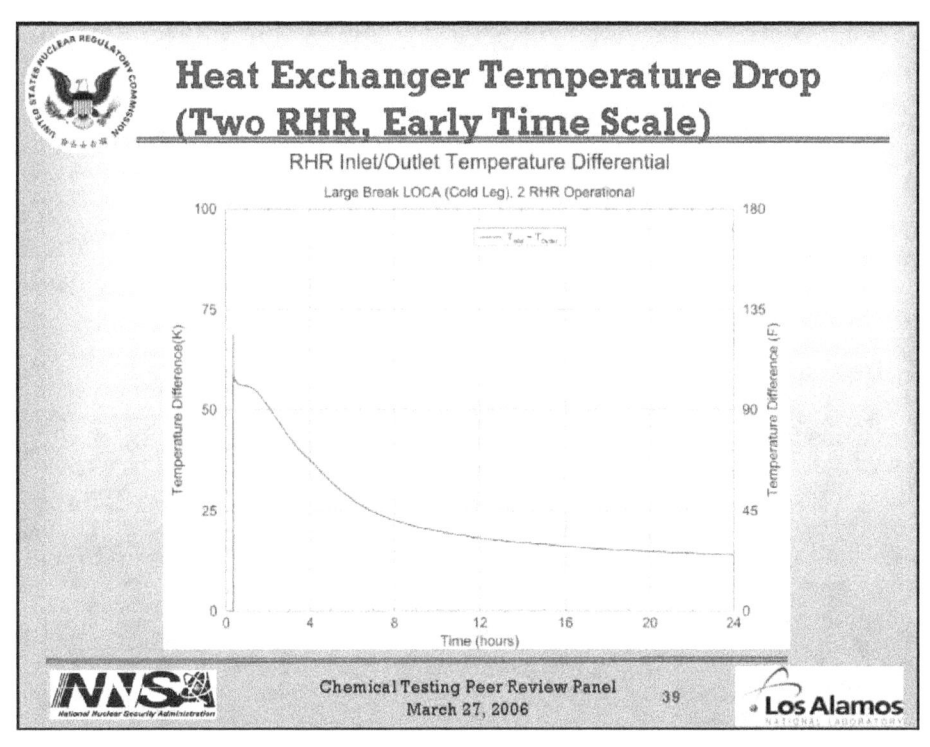
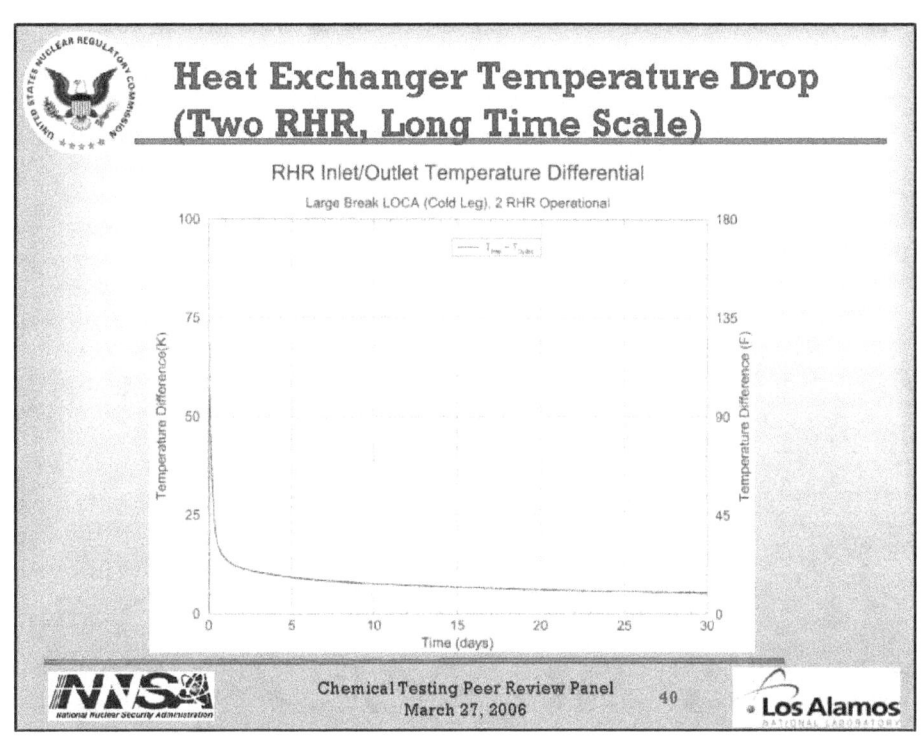

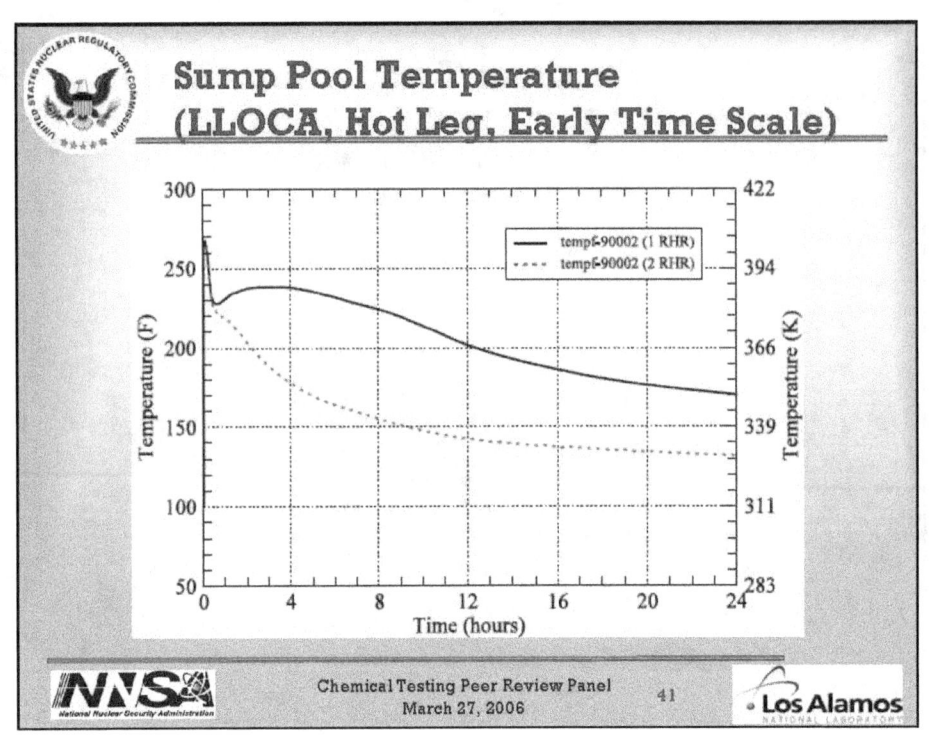
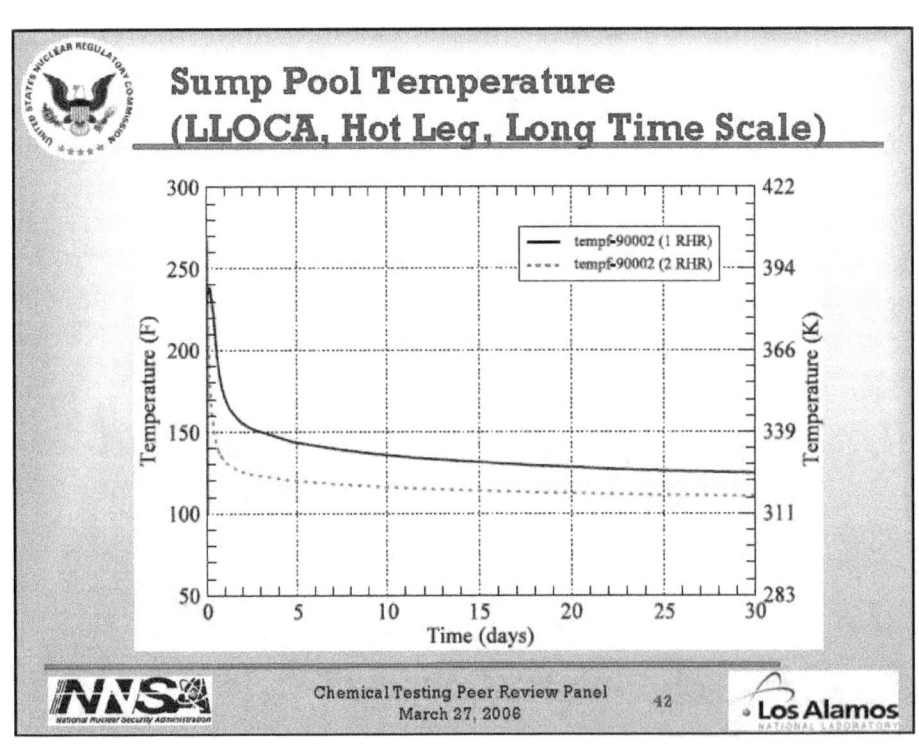

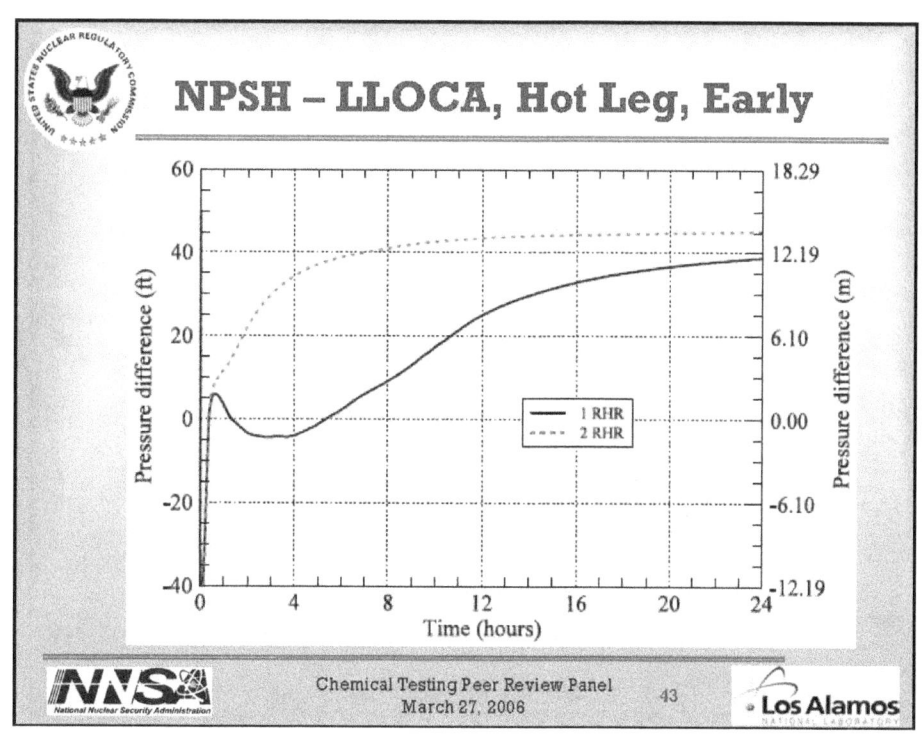
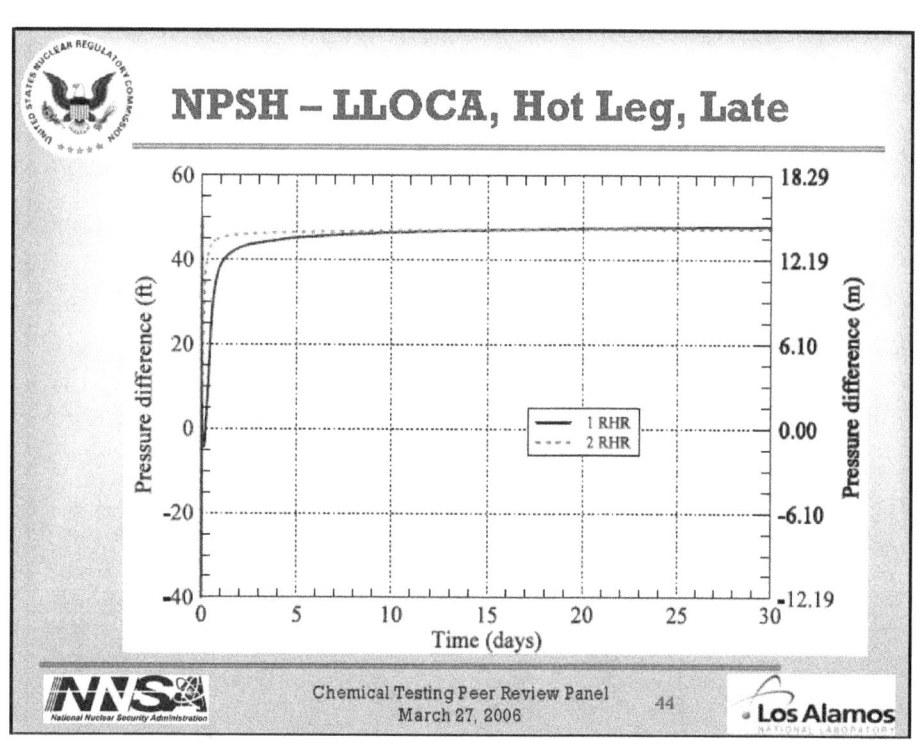

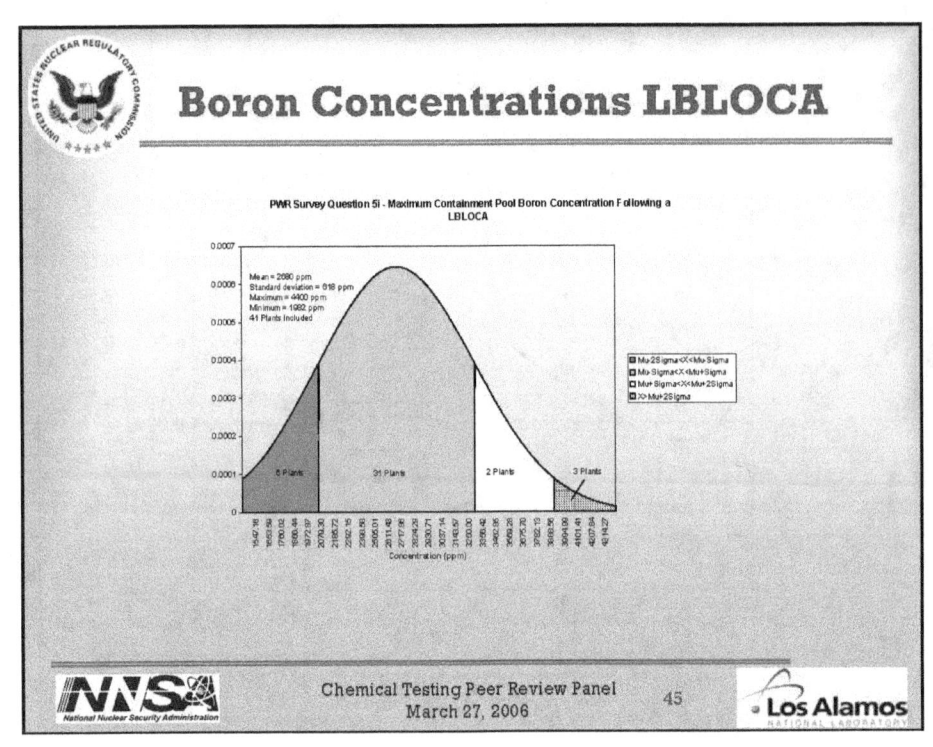
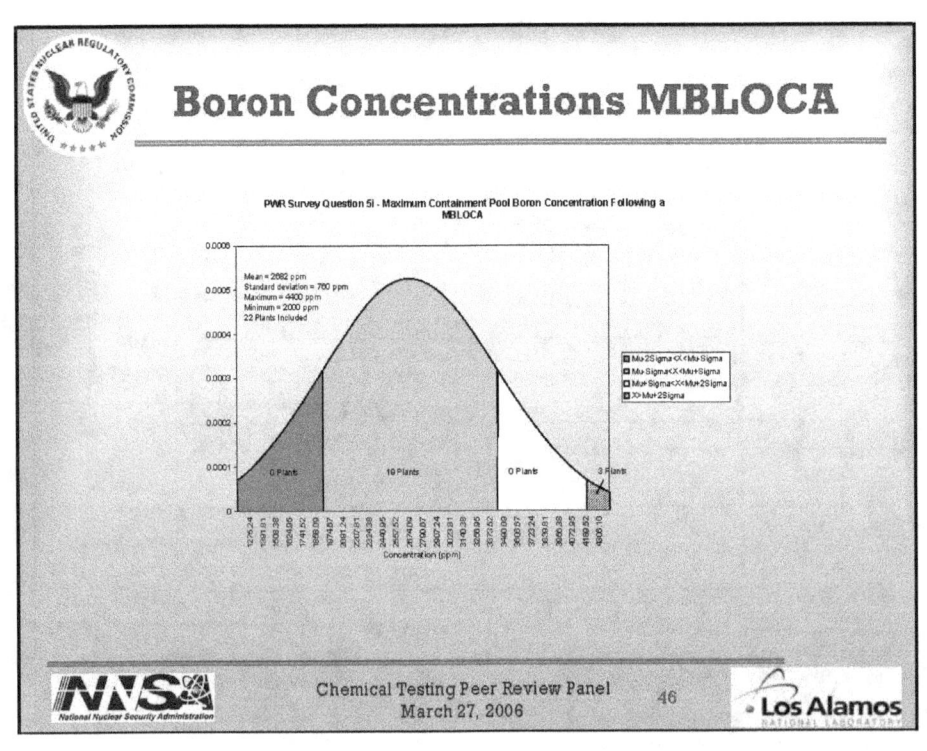

APPENDIX C

PIRT Issue Summary

Table C-1: Debris Generation (0 to 30 s)

Issue Number	Phenomena	Description	Implications
T1-1	Crud release	Fe, Ni corrosion oxides (≈ 125 μm layer) from the RCS piping are released due to hydraulic shock of the failure event	The crud release creates a radiolytic environment on materials caught on the sump screens which could affect subsequent reactions. Also, the oxides add particulate mass to containment pool, which may enhance coagulation.
T1-2	RCS coolant conditions at break	The RCS coolant chemistry varies over the fuel cycle. B concentrations $\approx$ 2000-4000 ppm (beginning) to 50 ppm (end of fuel cycle) B/Li ratio $\approx$ 100. The 2-phase jet is at temperatures of $\approx$ 315C to 120C upon cooling	Initial reactor water chemistry spewing out of the break and forming the containment pool will have variable B concentration. The Li/B ratio is maintained approximately constant.
T1-3	pH variability	pH @25C at beginning of fuel cycle is acidic (4) while closer to neutral (7) at end of cycle.	Similar implications to issue 2; variable reactor and initial containment pool chemical environment.
T1-4	Localized B concentration in jet	Droplet evaporation within the high temperature jet emanating from the pipe break forms at highly concentrated B aerosol solution. As boric acid concentrates at high temperatures (250C - 300C), a polymerization appears to occur that releases protons and causes the system to become very acidic.	1. Corrosion of impacting surfaces: enhanced erosion & corrosion vs. formation of a protective lacquer to limit erosion. 2. May limit B concentration in containment pool chemistry
T1-5	RCS fluid creates "oxidizing environment"	Coolant is released as a super heated spray once break occurs. Hydrogen (25 – 35 cc/kg of H_2 injected as gas) transfers to the gas phase.	1. H_2 evaporation creates a more oxidizing environment in the water phase by raising the redox potential. 2. The partial pressure of H_2 is lowered because of the large volume of the gas phase (containment volume) compared with the water volume being released. This also increases the redox potential.

Issue Number	Phenomena	Description	Implications
T1-6	Jet impingement	Water jet & fine debris within jet may impact surfaces and chip coatings	Initiates metallic pitting, corrosion and ablation of other materials (concrete, insulation)
T1-7	Break proximity to organic sources	RCP oil storage tank is made with ≈ ½" carbon steel with epoxy coating (250 gallon per pump, but tank may just contain leakage).	Tank failure timing: earlier in post-LOCA (jet impingement) vs. later in post-LOCA (corrosion) Organic sources to containment pool would alter chemistry (i.e., complexation, etc.)
T1-8	Break proximity to secondary systems	Break near cooling systems could result in failure of lines containing solvents (i.e., Freon) into containment environment	1. Radiolytic decomposition could create chlorides and fluorides which may impact corrosion and containment pool chemistry. 2. Possible effects on pump performance
T1-9	Debris mix particle/fiber ratio	Breaks in different locations will create different debris characteristics: total mass, mixture constituents, compositions	1. Could alter the containment pool chemistry. 2. Could affect debris bed formation on the sump screen and increase variability of chemical product capture efficiency.
T1-10	Hydrogen peroxide effects	Hydrogen peroxide will behave much the same as H_2O and most of it will remain in the liquid phase. Additional hydrogen peroxide evaporates into containment vapor space and may cause additional corrosion. It condenses on cooler surfaces and can accelerate corrosion there	1. Could affect how much H_2O_2 is in the containment pool: Evaporation & condensation could sequester form pool droplets could transport back to pool. 2. H_2O_2 alters containment pool chemistry and potentially causes changes in oxidation states or chemical forms in sump water..
T1-11	Nuclei Formation	Rapid subcooling and evaporation of water from effluent of broken pipe droplets might lead to precipitation of LiOH and other species. Concentrations increase and temperature decreases rapidly	1. Initial precipitate may be formed from species existing with in the RCS coolant. 2. Nuclei could form which could enhance precipitation of other species within the containment pool at a later time.

Table C-2: ECCS Injection (30 seconds to t_{recirc})[a,b,c]

Issue Number	Phenomena	Description	Implications
T2-1	Hydrogen sources within containment	H_2 concentrations in vapor and containment pool include the RCS inventory, Schikorr reaction, corrosion of metallic materials	Containment pool redox potential is a function of dissolved hydrogen which is established by the listed reactions. RCS evaporation may lead to equilibration with liquid, decreasing redox potential
T2-2	ECCS injection of Boron	After pipe break, B ($\approx$ 2800 ppm) is injected into RCS to cool reactor	Provides large B source which may affect chemical reaction products in containment pool. Specifically, the B source will serve as a pH buffer.
T2-3	Containment spray corrosion	Containment sprays for buffering creates wetting & surface films on material surfaces with water containing B, LiOH, H_2	Containment spray (up to 4 hours) could enhance corrosion formation on unsubmerged metallic species which could contribute to containment pool concentrations. Spray pH: NaOH = 8.5 – 10.5 TSP = 4.5 - 5 STB = 4.5 – 5
T2-4	NaOH pH control	NaOH injected through containment sprays.	Alkaline containment pool environment (pH = 8.5 – 10.5)
T2-5	STB pH control	Released with ice in containment during a LOCA	Alkaline containment pool environment. Initially pH is high 9 initially and then decreases with mixing (pH = 8 – 9)
T2-6	TSP pH control	Contained in basket on containment floor that dissolve	Slightly alkaline containment pool environment. Initially pH is high (pH = 11) until complete mixing within about 4 hours (pH = 7 – 8)
T2-7	Containment spray transport	Sprays wash latent debris, corrosion products, insulation materials, & coating debris into containment pool	Affect on containment debris sources (types, amounts, compositions) and contributions to the chemical sump pool environment.
T2-8	Containment spray CO_2 scavenging	Containment sprays cause CO_2 absorption within containment pool and carbonate formation.	Affect on containment pool concentrations of CO_2
T2-9	Debris dissolution begins	Debris dissolution begins. Initial expected products include Cal-Sil, cement dust, organic fiberglass binders, epoxy & alkyd coatings, uncoated concrete	Indicate potential important contributors (if any) to chemical containment pool environment during this time frame. Dissolution of other products will occur over longer time frames.

Issue Number	Phenomena	Description	Implications
T2-10	Carbonate concentration	Carbonates in solution are provided by atmospheric CO_2, and concrete dust at this point	1. Create nucleation sites for species precipitation 2. scavenge dissolved Ca to impede formation of other compounds to form calcium carbonate
T2-11	Containment pool mixing	High velocity, turbulent water flow exists underneath the pipe break location, remainder of the pool relatively quiescent	Mixing of aqueous/gas phases at the break location is enhanced; could aid CO_2 absorption; buffering agent dissolution/mixing, liquid/vapor exchange of species.
T2-12	Boric acid corrosion of exposed concrete	Concentrated boric acid solutions in contact with exposed concrete	1. Calcium boro-silicates may precipitate 2. pH decrease in containment pool possible
T2-13	Fe, Ni radiological reaction	Initial Fe, Ni oxide breaks off due to TH transient. Reduced Ni, Fe is dissolved in RCS, combines with air, oxidizes and forms hematite, maghemite, and magnetite. . Kilorad/hour and higher rad. dose rates arise throughout containment due to dispersal of Fe and Ni activation products.	1. Solid particulate is added to containment pool, another debris source 2. High pH will increase the reaction. 3. Reaction continues until RCS piping cools down (1 day or more). 4. Radiolysis products generated throughout post-LOCA cooling system
T2-14	Hydrolysis	Nickel oxide becomes catalyst for producing H_2 from radiolysis	Affects redox potential
T2-15	Organic Complexation	Organic acids become absorbed on surfaces of solids & inhibit solid species growth; example: aliphatic acid	Effectively enhances solubility limits. Solid species may precipitate, but remain relatively small in size (nano-scale). They do not agglomerate or grow to macroscopic sizes.
T2-16	Organic Sequestration	Organic electron-rich atoms combine with soluble metal compounds.	Sequesters metal ions so that solid species precipitation is delayed or does not occur.
T2-17	Auxiliary Component Cooling Line Failure	LOCA causes attached or nearby component cooling line failure which could release a number of chemicals into the containment pool (chromates, molybdates (6000 – 1000 ppm in storage tank), nitrites, tolytriazole (25 ppm in storage tank), benzotriazole, hydrazine	1. The release of these types of chemicals could increase the ionic strength of the containment environment which may decrease gel stability. 2. Chemicals may accelerate corrosion.

Issue Number	Phenomena	Description	Implications
T2-18	Polymerization	Precursor to precipitation & agglomeration; metals/oxygen bonds, ripen to form covalent bonds, growth until they qualify as ~nanometer particles size: Si, Al, Fe, Boric acid are candidates	May be necessary to form large enough particles to result in tangible effects on ECCS performance
T2-19	Co-precipitation	Method of precipitation/separation Examples: Ni/Fe/Cr, Al/Si/B; Co/Fe systems; Precipitation of one species leads to precipitation of other species (below solubility limit); Radioactive elements precipitate (at low concentrations), activation products: ex. Strontium, Ni, Silver, Fe, Co, Zr	More solids species form which could lead to greater concentration of chemical products at the sump screen or downstream.
T2-20	Radiolytic environment	Radiolysis reaction creates an oxidizing environment and can change solution's electric potential; H_2, O_2, and H_2O_2 balance most important for determining redox potential	Could affect chemistry in containment pool which affects species which form (e.g., Hanford tank: mixture of different components @ different phases)
T2-21	Inorganic Agglomeration	Formation of larger clumps of smaller particulates: Depends on PZC; ionic strength (the higher the strength the smaller the distance for agglomeration)	1. May be necessary to form large enough particles to result in tangible effects on ECCS performance 2. Can occur quickly if conditions are right. 3. Existence of organic species can increase or decrease likelihood (See T2-24)
T2-22	Galvanic Effects	Electrical contact between metals results in increased corrosion of less noble metals: example: magnetite precipitation on Aluminum	1. Increases in the corrosion source term for less noble (anodic) metal. 2. Decreases in corrosion source term for more noble (cathodic) metal.
T2-23	Deposition & Settling	Possibility that chemical products formed during this time period either settle within containment pools or are deposited on other surfaces	Potentially remove substantial quantities of material from transporting to the containment sump screen: indicate expected species that are most likely to settle, if possible.

Issue Number	Phenomena	Description	Implications
T2-24	Organic Agglomeration	Formation of larger clumps of smaller inorganic particulates nucleating around organic acids or oil (such as soap coagulates dirt particles)	Coagulated particles can collect on sump screen, decreasing flow, or collect in other places to decrease the loading on the sump screen

[a] Massive fuel damage is not considered during this time period because the objective is to prevent this through successful operation of the ECCS system. Radiological and other contributions to the chemical environment from minor amounts of fuel damage are expected to be insignificant compared to crud release and other phenomena listed in these tables.
[b] Sump pool temperatures can be as high as 120C initially. After 30 minutes, between 60 – 90C
[c] Phenomena listed in earlier time periods (i.e., T1) may continue to be active or important during this time.

Table C-3: ECCS Recirculation (t_{recirc} to 24 hours)[a,b]

Issue Number	Phenomena	Description	Implications
T3-1	TSP pH control	Contained in basket on containment floor that dissolve	Slightly alkaline containment pool environment (pH = 7 – 8); all TSP dissolved within 1 to 4 hours post LOCA.
T3-2	NaOH pH control	NaOH injected through containment sprays.	Alkaline containment pool environment (pH = 8.5 – 10.5); pH buffering completed typically within 4 hours.
T3-3	STB pH control	Released with ice in containment during a LOCA	Alkaline containment pool environment (pH = 8 – 9); all STB should be introduced within a few hours.
T3-4	NaOH Injection	NaOH, B laden spray impinges on material surfaces	1. Solution may accelerate corrosion of metallic/non-metallic surfaces due to high pH, boron concentration. 2. Solution may inhibit corrosion by formation of an initial passive layer impervious to future corrosion.
T3-5	Cable degradation	The insulation surrounding electrical cables located throughout containment is affected by radiolysis, releasing chlorides.	The chlorides affect radiolysis and ionic strength of containment pool environment.
T3-6	Radiolytic environment	Radiolysis reaction changes solution's redox potential; H_2, O_2, and H_2O_2 balance; peroxide formation occurring during this time frame.	Could affect chemistry in containment pool which affects species which form - most important for determining redox potential (e.g., Hanford tank: mixture of different components @ different phases)
T3-7	Fiberglass leaching	Fiberglass-based insulation products begin leaching constituents; e.g., Si, Al, Mg, Ca, etc.	Dissolution source term for producing ions which contribute to containment pool chemistry.
T3-8	Secondary system Contamination	SG tube rupture leads to secondary side chemicals being added to containment pool.	1. Possible additional source for organic materials which could delay precipitation through complexation or by sequestering metal ions. 2. Could accelerate material corrosion.
T3-9	Flow-induced nucleation[c]	Nucleation sites created from cavitation, deareation, air entrainment, turbulence (attachment opportunity)	Increases possibility of precipitation/chemical species formation

Issue Number	Phenomena	Description	Implications
T3-10	Turbulent mixing[c]	Turbulence in containment pool (such as directly under the break) either inhibits or promotes precipitate formation and growth	1. Could inhibit growth due to turbulent shear forces which don't allow solid species to agglomerate. 2. Could promote formation and growth by providing more opportunity for particles to collide and overcome diffusion. 3. Affects the solid species concentration within the containment pool which could contribute to head loss & other ECCS performance degradation.
T3-11	Quiescent settling of precipitate[c]	Quiescent flow regions within containment pool promote settling	Little flow allows larger size, more stable particles/precipitates to form which promotes settling.
T3-12	Electrostatic scavenging	Material surfaces electrostatically attract particles and lead to deposition/agglomeration on the surface; thin accumulation limited to 2 to 3 layers of material. It is a function of solution PZC and surface and particle charges.	1. Provides a mechanism for retaining chemical products either in benign areas (e.g., walls, surfaces) or in areas that may impact system performance (e.g., sump screen, pipe internals). 2. Surface layer either inhibits or enhances material corrosion.
T3-13	Chemically induced settling	Chemical species attach to or coat particulate debris which leads to settling. Examples are Al coating on Nukon fiber shifting the PZC, or formation of a hydrophobic organic coating	Results in less particulate debris and chemical product transporting to and either accumulating on or passing through the sump screen.
T3-14	Agglomeration & Coagulation	Formation of larger clumps of smaller particulates: Depends on PZC; ionic strength (the higher the strength the smaller the distance for agglomeration); sensitive to many factors including shape factors, maximum particle size.	1. May be necessary to form large enough particles to result in tangible effects on ECCS performance 2. Can occur quickly if conditions are right. 3. Existence of organic species can decrease likelihood
T3-15	Particulate nucleation sites	Particles within containment create nucleation sites for chemical precipitation: examples include radiation tracks; dirt particles; coating debris; insulation debris; biological debris, etc.	Environment is created which foster formation of solid species that could lead to ECCS degradation.

Issue Number	Phenomena	Description	Implications
T3-16	Additional debris bed chemical reactions	Concentration of radionuclides (100's of Curies available) acts as a "resin bed" or chemical reactor. A number of possible reactions occur.	1. Hydrogen peroxide formation 2. $Fe^{+2} \rightarrow Fe^{+3}$ increase due to increase in redox potential; 3. Organic materials decompose 4. possible coating of NUKON to reduce solubility; glass embrittlement (shorter fiber);
T3-17	Sump screen: high localized chemical concentrations	High localized concentrations of certain species drive the reaction efficiency at the sump screen	1. Effect could be at the exterior surface of insoluble materials, the internal interstitial void material, or both 2. Possible effect on the chemical species which form within the sump screen bed.
T3-18	Sump screen: fiberglass morphology	Localized chemistry alters the fiberglass contribution to the chemical environment	1. Possible coating of fiberglass fibers reduces leaching/solubility. 2. Radiation leads to glass embrittlement; shorter fibers form which can pack more densely (increase head loss) and provide additional surface area for leaching/reactions.
T3-19	ECCS pump: seal abrasion	Abrasive wearing of pump seals (e.g., magnetite - high volume/concentration of mild abrasive) creates additional materials that contribute to sump pool chemistry.	1. Additional particles that may contribute to reactor core clogging 2. Particles may add additional sump screen loading. 3. Particles may affect chemical species formation.
T3-20	Heat exchanger: solid species formation	Concentrations/species that are soluble at the containment pool temperature precipitate at lower ($\Delta T \sim 15$-$20C$) heat exchanger outlet temperature.	1. Species remain insoluble at higher reactor temperatures and affect ability to cool the reactor core. 2. Species remain insoluble at higher containment pool temperatures and cause additional head loss upon recirculation.
T3-21	Heat exchanger: deposition & clogging	Solid species which form in the heat exchanger lead to surface deposition and/or clogging within closs-packed head exchanger tubes (5/8" diam.)	1. Severe clogging/deposition causes flow decrease through heat exchanger core and an inability to cool reactor core. 2. Less severe deposition may degrade heat transfer and degrade heat flow from the reactor core.

Issue Number	Phenomena	Description	Implications
T3-22	Reactor core: fuel deposition	ΔT increase (+70C from sump pool) and retrograde solubility of some species (e.g., Ca silicate, Ca carbonate, zeolite, sodium calcium aluminate) causes scale, build-up on reactor core	1. Decreases heat conduction values from fuel. 2. Localized boiling occurs due to insufficient heat removal. 3. Deposits spall off, creating additional debris source.
T3-23	Reactor core: hydrogen increases	H_2 builds up in from cladding oxidation	1. Affect redox potential and chemical environment. 2. Increases containment hydrogen concentration; may lead to conflagration.
T3-24	Reactor core: diminished heat transfer	Insulation debris and chemical products mixed within water cause a reduction in the effective heat transfer capabilities (C_p) of mixture. The precipitation of this material is initiated in RHR heat exchanger.	Ability to remove heat from the fuel is diminished.
T3-25	Reactor Core: Blocking of flow passages	Chemical products, in combination with other debris, blocks primary flow passages for getting cooling water through core	Heat transfer may be impeded; flow is forced to bypass the lower plenum debris screens
T3-26	Reactor Core: Particulate settling	Particulate settling occurs due to relatively low, upwards flow (for cold leg injection) within reactor	Compacted deposits forms which may impede heat transfer and water flow, especially for lower portions of reactor fuel.
T3-27	Reactor Core: Precipitation	ΔT increase (+70C from sump pool) and retrograde solubility of some species (e.g., Ca silicate, Ca carbonate, zeolite, sodium calcium aluminate) causes precipitation, additional chemical product formation	1. Products contribute to debris in reactor core which blocks flow passages, impedes heat transfer. 2. Additional precipitate is created which travels to the sump screen and leads to head loss.
T3-28	Exposed, Uncoated Concrete Dissolution	Dissolution can result in pH increases (depending on surface area) up to 10.5 below concrete surface relatively quickly.	1. This could result in pH increase in containment pool over time. 2. Leaching of concrete elements (e.g., Ca, Si, etc.) into containment pool.
T3-29	Coatings Dissolution	Dissolution/leaching of epoxy, alkyd, or zinc-based coatings and primers occurs	1. Additional source term for products affecting the containment pool chemistry. 2. Chlorides affect radiolysis and ionic strength of containment pool environment.

Issue Number	Phenomena	Description	Implications
T3-30	Boric Acid Corrosion	Concentrated boric acid pooled on unsubmerged portions of large ferritic steel components (e.g., piping RPV, pressurizer, steam generator) continues to cause corrosion of these materials.	1. Contribute additional Fe ions which may be transported to containment pool. 2. Could affect coprecipitation and solid species concentrations 3. May lead to long-term structural weakness if severe.
T3-31	CO_2/Carbonate Radiolysis	Radiolysis of CO_2 and carbonates could lead to the formation of organic acid ligands (e.g., acetate, butyrate, oxalate) This is observed during the leaching of spent fuel by groundwaters containing bicarbonate ions	1. Phenomena would complex metal species in solution, thereby enhancing dissolution rates and diminishing the precipitation of phases containing these complexed species. 2. Could occur within containment pool, reactor vessel, or within the sump screen debris bed.

[a] Sump pool temperatures are expected to range between 50 – 90C during this time period.
[b] Phenomena listed in earlier time periods (i.e., T1, T2) may continue to be active or important during this time.
[c] Bulk flow velocities = 0.005 ft/s to 0.1 ft/s (0.15 – 3 cm/s)

Table C-4: ECCS Recirculation (24 hours to 15 days) [a,b]

Issue Number	Phenomena	Description	Implications
T4-1	Source terms: Unsubmerged materials	Corrosion of various materials (Al, Fe, concrete, Cu, Zn) due to condensation and transport into containment pool; condensation expected due to non-uniform containment temperature	1. Contributes species to containment pool 2. Condensate chemistry & environment governs corrosion/leaching rates.
T4-2	Submerged source terms: Pb shielding	Any acetates present in containment pool will dissolve Pb, which could lead to formation of lead carbonate particulate. Lead blanketing to shield hot spots and covered with plastic coating, but coating likely destroyed. Several hundred pounds of lead in flat sheets. Some plants still use lead wool. No Pb in ICET program.	Would provide additional particulate loading within containment pool
T4-3	Submerged source terms: Cu	Concern stems not from Cu compounds, but the various effects that Cu may have on other corrosion processes. Cu concentrations evaluated in ICET program; Cu comes from containment air coolers, motor windings and grounding straps.	1. By forming a galvanic couple, can facilitate attack of other metals (e.g., Al). 2. Cu ion deposition can occur which may inhibit corrosion. 3. Within an oxygenated environment, Cu can accelerate corrosion
T4-4	Submerged source terms: Fe	Boron inhibits pitting corrosion on surface of steel structures. Fe concentrations evaluated in ICET program.	1. Less Fe ions in containment pool. 2. less corrosion and structural weakening of steel structures
T4-5	Submerged source terms: Al, decreased concentrations	Less dissolved Al, than in ICET #1, affects the type and quantity of chemical byproducts that form at high pH.	1. Less corrosion inhibition of insulation results in greater Si levels. 2. Less Al corrosion products allows other species to form within pool
T4-6	Submerged source terms: Al*, increased concentrations	Specific mix of other metals in containment pool leads to more Al corrosion (catalytic effect) at either lower or higher pH	Increases Al solid species which may form.

Issue Number	Phenomena	Description	Implications
T4-7	Submerged source terms: Fiberglass dissolution	Dissolution at high pH contributes Silicates, Cu, Na, Al, Mg, B, misc. organics to containment pool. Fiberglass concentrations in ICET test plan	1. Solid species formation is affected by silicates, Cu, Na, Al, Mg, B concentrations. 2. Misc. organics in sump pool affect complexation.
T4-8	Submerged source terms: Fiberglass inhibition	Leaching of fiberglass inhibits corrosion of certain metals (e.g., Al)	Less metallic ions within containment pool
T4-9	Submerged source terms: Zn passivation	Zn passivation decreases corrosion passivation of Fe which allows Fe corrosion.	Galvanized steal/zinc coating; vessel site for ...; dissolve ...; precipitate as silicate; ZnO, ZnOH corrosion products;
T4-10	Submerged source terms: Zn corrosion products	Zn oxide and hydroxide corrosion products form which contributes to solid species loading.	Additional source term that could contribute to sump screen clogging and/or downstream effects.
T4-11	Submerged source terms: Zn coprecipitation	Zn may coprecipitate with other species (, notably Fe, Al). Reactor Vessel a possible site for coprecipitation	Additional source term that could contribute to sump screen clogging and/or downstream effects.
T4-12	Zinc hydroxide dissolution (zincate)	Buffered pH containment pool dissolves zinc hydroxide so that other species, particularly silicates, may precipitate	Zinc silicates may be more likely to contribute (more transportable, possibly amorphous) to head loss or downstream effects.
T4-13	Submerged source terms: Zn-based coatings	Leaching of Zn-based primers (primarily inorganic Zn phosphates) creates possibility for additional dissolved Zn in solution	Lead to formation of zinc oxide/hydroxides, carbonates, or silicates that could contribute to head loss/downstream effects.
T4-14	Source Term: Seal table corrosion	Multiple materials in seal table (Inconel, SS, plastics, organics) leads to formation of additional dissolved species. Seal table is an area in bottom of core where instruments are inserted into the core. A series of pressure isolation chambers. Generally lower than bottom of containment floor	Additional dissolved species present within the containment pool

Issue Number	Phenomena	Description	Implications
T4-15	Submerged source terms: Fire barriers	Silicon-based seals (bisco seals) used at many structural penetration points at fire barriers. These may leach in post-LOCA environment	Another silicon source term that could contribute to formation of chemical byproducts which could impact head loss and downstream performance.
T4-16	Submerged source terms: organic buoyancy	Organics coat materials (e.g., aluminum, Cal-Sil, fiberglass, etc.) and increase the buoyancy of particulate so that it's more likely to float	This could enhance transport of particulates or chemical byproducts to the sump screen.
T4-17	Submerged source terms: coatings	Leaching of submerged coatings to contribute species to the containment pool. Possible sources include Pb-based paints (older containment buildings), phenolics, PVC	Additional source terms (especially chlorides, fluorides) which contribute to containment pool chemistry.
T4-18	Submerged source terms: RCP oil tank failure	Overflow, failure, or leakage occurs either by LOCA or preexisting condition. Oils and other organics are released into containment pool.	Oils and other organics may affect complexation and sequestration of metallic species. Containment pool chemistry and byproduct production are influenced.
T4-19	Submerged source terms: biofilm formation	Biofilms form which protect against metallic corrosion (by forming passive layer) or lead to production of acids which increase metallic corrosion. Some biofilms can be preexisting in containment from outages.	1. If decreases metallic corrosion, would reduce containment pool concentration of certain dissolved species. 2. If increases metallic corrosion, would increase containment pool concentration of certain dissolved species. 3. May eventually lead to SCC failures of components.
T4-20	Submerged source terms: biologically enhanced corrosion	General corrosion is enhanced due to biological agents. Examples include polysaccharides and sulfate reducing bacteria which may enhance Fe corrosion. Sulfur or carbon sources contribute to bacteria formation.	Fe and other metallic corrosion is enhanced leading to increased concentrations of metallic species within the containment pool.

Issue Number	Phenomena	Description	Implications
T4-21	Submerged source terms: biologically enhanced H_2 embrittlement	Anaerobic and aerobic bacteria synergistically enhance stress corrosion cracking and hydrogen embrittlement of steels. Sulfur or carbon sources contribute to bacteria formation.	Structural cracking enhanced which could weaken and lead to failure of structural components.
T4-22	Submerged source terms: biological growth in debris beds	Bacteria grow in preexisting debris beds located on the sump strainer screen or elsewhere within the ECCS system.	Increased source term which can contribute to clogging or detrimental performance of pumps, valves, etc.
T4-23	Submerged source terms: electrical insulation	Chlorides form due to radiological breakdown of cable insulation. This affect was simulated in the ICET experiment.	1. Additional source term for products affecting the containment pool chemistry. 2. Chlorides affect radiolysis and ionic strength of containment pool environment.
T4-24	Reactor Core: Fuel spalling	Zr-oxide layer spalls off reactor fuel due to temp gradient - 50 micron layers exist	1. Heat transfer depends on balance of oxide growth, deposition, and particulates, particulates may also enhance coagulation in core that may diminish heat transfer of water. 2. Additional solid products which contribute to clogging within the reactor core. 3. Additional products which could lead to clogging within the reactor core or be transported back to the sump screen.
T4-25	Radiological** effects: debris bed accumulation	Radio nuclides transport to, accumulate, and become concentrated within the sump screen debris bed altering the local chemical conditions	1. More concentrated effect than if spread throughout containment pool 2. Directly alter chemical byproducts formed at the sump screen, contributing to head loss.
T4-26	Radiological** effects: Dissolution & oxidation changes	Radiolysis leads to changes in the oxidation state in reactor, piping, and containment materials which affects the dissolution rate and the byproducts which form.	Affects the chemical byproduct source term loading that is present in the containment pool.

Issue Number	Phenomena	Description	Implications
T4-27	Radiological** effects: Radiolytic affect on biofilms	Radiolysis inhibits the formation and growth of biofilms	Films may be either advantageous or detrimental to corrosion and chemical byproduct formation (see earlier items)
T4-28	Radiological** effects: Redox potential changes	Radiolysis causes changes in the redox potential either in the containment pool or in debris beds (on the sump screen, or within ECCS component)	Could affect chemistry in containment pool which affects species which form (e.g., Hanford tank: mixture of different components @ different phases)
T4-29	Radiological** effects: corrosion rate changes	Low doses and low temperature elevate the corrosion rate through formation of hypochlorite through radiolysis of Cl bearing water. Could increase corrosion rates (esp. pitting corrosion) of Al, SS, Fe.	Corrosion rate increases could increase the amount of metallic and nonmetallic species in containment; alters chemical byproduct formation.
T4-30	Radiological** effects: agglomeration	Radiolysis enhances agglomeration of chemical species.	Advantage: Agglomeration could lead to increase settling rate of product. Disadvantage: Could form products that are more likely to cause clogging of sump screens, downstream components.
T4-31	CO_2/O_2 air exchange	Air exchange provides source for CO_2 ingestion within the containment pool. CO_2 quantities limited by containment volume. Radiolysis can also promote formation of carbonates.	Increase solid species in containment pool.
T4-32	ECCS Pumps: erosion/corrosion	Chemical byproducts cause erosion or corrosion of pump internals, especially tight tolerance components (bearings, wear rings, impellers, etc.).	Pump performance degrades, possibly to the point of being inoperable.
T4-33	ECCS Pumps: seal degradation	Chemical environment causes leaching/degradation of pump seal materials or chemical byproducts cause seal erosion	Pump performance degrades, possibly to the point of being inoperable.

Issue Number	Phenomena	Description	Implications
T4-34	Heat exchanger: secondary contaminants	Leaks in heat exchanger tubes allow secondary-side water additives to migrate over to the containment pool.	Additives in the secondary side water chemistry alter the chemistry in the containment pool.
T4-35	Heat exchanger: precipitate formation	Drop in temperature leads to the formation of sold species (e.g., AlOOH, FeOOH amorphous SiO_2) and ripening such that macroscale coatings and/or suspended particulates form.	1. Create clogging, and less efficient heat transfer for cooling the ECCS recirculating water. 2. Form products that may clog reactor core and degrade heat transfer from fuel. 3. Products form which transport back to sump screen and lead to head loss. 4. Particulates act as nucleation sites for other compounds to precipitate
T4-36	pH drop in containment pool	pH drop due to evolving containment pool chemistry causes Al (and other metallic species) to precipitate.	Creates solid particulate loading that could affect sump screen head loss, heat exchanger clogging, reactor core clogging, and additional nucleation sites.
T4-37	Reactor Core: continued deposition/precipitation	Zn, Ca, Mg, CO_2 based deposits, films, and precipitates may form at higher temperatures within the reactor core	1. Diminished heat transfer from the reactor fuel. 2. Spalling could create additional solid products which contribute to clogging within the reactor core. 3. Spalling could create additional solid products which contribute to sump screen head loss.
T4-38	Reactor Core: fuel deposition spalling	Zn, Ca, Mg, CO_2 based deposits and films which form on the reactor core spall	1. Additional solid products which contribute to clogging within the reactor core. 2. Additional solid products which contribute to sump screen head loss.
T4-39	Transport Phenomena: amorphous coating	Amorphous silica forms on surface of chemical precipitates (e.g., AlOOH, FeOOH). Coatings increase density and make it less likely for products to transport to sump screen.	Decreased transportability will result in less product migrating to or through the sump screen.
T4-40	Transport Phenomena: precipitation/co-precipitation	Precipitation/co-precipitation and ripening of solid species within containment pool creates solid species which are less likely to transport.	Decreased transportability will result in less product migrating to or through the sump screen.

Issue Number	Phenomena	Description	Implications
T4-41	Transport Phenomena: Metallic scouring	Turbulent flow causes chemical films/products to be scour off the surface of metallic components so that they enter the containment pool.	Increased particulate loading that may increase head loss, degrade pump, valve, heat exchanger performance, and/or affect heat transfer from reactor fuel.

[a] Sump pool temperatures are expected to range between 50 – 70C during this time period.
[b] Phenomena listed in earlier time periods (i.e., T1, T2, T3) may continue to be active or important during this time.
[c] Al, in ICET tests, has provided an important contribution to observed chemical effects both through precipitate formation at high concentrations and pH and in inhibiting fiberglass leaching.
[d] Soluble/insoluble activation species include, for example, Na^{24}, Be^{7}, Cr^{51}, $Co^{58,60}$, $Mn^{54,56}$, $Zr^{95,97}$, $Fe^{55,59}$, $Ni^{59,63}$, $Nb^{95,97}$, Sb^{125}, Zn^{65}, $Cs^{134,137}$, and $I^{131,133}$

Table C-5: ECCS Recirculation (15 days to 30 days) [a,b]

Issue Number	Phenomena	Description	Implications
T5-1	Organic radiolysis	Radiolysis of organics leads to carbonate formation	1. Create nucleation sites for species precipitation 2. scavenge dissolved Ca to impede formation of other compounds 3. Additional solid species contribution which leads to degraded system performance (head loss, etc.)

[a] Sump pool temperatures are expected to range between 40 – 60C during this time period.
[b] Phenomena listed in earlier time periods (i.e., T1, T2, T3, T4) may continue to be active or important during this time.

Definitions:
ECCS: Emergency Core Cooling System
RCS: Reactor Coolant System
Crud: Ni, Fe based oxides that reside on internal RCS piping.
RCP: Reactor Coolant Pump
LOCA: Loss-of-Coolant-Accident
t_{recirc}: Time at which ECCS recirculation pumps are actuated
TSP: Trisodium phosphate buffering agent {$Na_3(PO_4)$}
STB: Sodium tetraborate buffering agent {$Na_2B_4O_7$}
Cal-Sil: calcium silicate insulation
TH: Thermohydraulic
PZC: pH of Zero Charge
In: Inconel (nickel-based alloy)
SS: stainless steel
RHR: Residual Heat Removal System (also referred to as Decay heat removal system)

APPENDIX D

PIRT RESPONSE TABLES

Table D-1: Implications and Time Table

Location	Phenomena/Implication Category	Time Period (t = 0 @ pipe break, beginning of LOCA)				
		0 – 30s	30s - t_{recirc}	t_{recirc} – 24h	24h – 15d	15d – 30d
Containment Pool	Break Location & System Failures					
	Reactor Chemistry Conditions					
	pH Effects					
	Radiological Effects					
	Organic Effects					
	Redox Potential & Effects					
	Carbonate Formation					
	Unsubmerged Material Corrosion					
	Submerged Metal Corrosion					
	Submerged Material Degradation					
	Precipitate Formation & Growth					
	Mixing & Flow Effects					
	Settling & Deposition					
Sump Screen	Debris Bed Buildup & Head Loss					

	Debris Bed Compaction & Head Loss					
	Debris Bed Reduction & Screen By-Pass					
	Chemical Species Generation or Reduction					
ECCS Pumps	Seal Degradation					
	Internal Component Erosion/Corrosion					
	Chemical Species Generation					
Heat Exchanger	Heat Transfer Reduction: Clogging or Deposition					
	Chemical Species Generation or Reduction					
Reactor Core	Heat Transfer Reduction: Clogging, Deposition, or Fluid Conditions					
	In-Core Settling					
	Chemical Species Generation or Reduction					

Table D-2: Importance Ranking for T1 Phenomena

Phenomenon	Importance* (H, M, L)	Rationale	Knowledge** (K, PK, UK)	Rationale
T1-1				
T1-2				
T1-3				
T1-4				
T1-5				
T1-6				
T1-7				
T1-8				
T1-9				
T1-10				
T1-11				

* L = Low, M = Medium, H = High
** K = Known, PK = Partially Known, UK = Unknown

Table D-3: Importance Ranking for T2 Phenomena

Phenomenon	Importance* (H, M, L)	Rationale	Knowledge** (K, PK, UK)	Rationale
T2-1				
T2-2				
T2-3				
T2-4				
T2-5				
T2-6				
T2-7				
T2-8				
T2-9				
T2-10				
T2-11				
T2-12				

T2-13				
T2-14				
T2-15				
T2-16				
T2-17				
T2-18				
T2-19				
T2-20				
T2-21				
T2-22				
T2-23				
T2-24				

* L = Low, M = Medium, H = High
** K = Known, PK = Partially Known, UK = Unknown

Table D-4: Importance Ranking for T3 Phenomena

Phenomenon	Importance* (H, M, L)	Rationale	Knowledge** (K, PK, UK)	Rationale
T3-1				
T3-2				
T3-3				
T3-4				
T3-5				
T3-6				
T3-7				
T3-8				
T3-9				
T3-10				
T3-11				
T3-12				

T3-13				
T3-14				
T3-15				
T3-16				
T3-17				
T3-18				
T3-19				
T3-20				
T3-21				
T3-22				
T3-23				
T3-24				
T3-25				
T3-26				

T3-27				
T3-28				
T3-29				
T3-30				
T3-31				

* L = Low, M = Medium, H = High
** K = Known, PK = Partially Known, UK = Unknown

Table D-5: Importance Ranking for T4 Phenomena

Phenomenon	Importance* (H, M, L)	Rationale	Knowledge** (K, PK, UK)	Rationale
T4-1				
T4-2				
T4-3				
T4-4				
T4-5				
T4-6				
T4-7				
T4-8				
T4-9				
T4-10				
T4-11				
T4-12				

T4-13				
T4-14				
T4-15				
T4-16				
T4-17				
T4-18				
T4-19				
T4-20				
T4-21				
T4-22				
T4-23				
T4-24				
T4-25				
T4-26				

T4-27				
T4-28				
T4-29				
T4-30				
T4-31				
T4-32				
T4-33				
T4-34				
T4-35				
T4-36				
T4-37				
T4-38				
T4-39				
T4-40				

T4-41				

* L = Low, M = Medium, H = High
** K = Known, PK = Partially Known, UK = Unknown

Table D-6: Importance Ranking for T5 Phenomena

Phenomenon	Importance* (H, M, L)	Rationale	Knowledge** (K, PK, UK)	Rationale
T5-1				

* L = Low, M = Medium, H = High
** K = Known, PK = Partially Known, UK = Unknown

APPENDIX E

PIRT Panelist Evaluations of T1 Phenomena:

Debris Generation from Pipe Break to 30s into LOCA

Table E-1: Debris Generation (0 to 30 s)

Issue Number	Phenomena	Description	Implications
T1-1	Crud release	Fe, Ni corrosion oxides ($\approx$ 125 μm layer) from the RCS piping are released due to hydraulic shock of the failure event	The crud release creates a radiolytic environment on materials caught on the sump screens which could affect subsequent reactions. Also, the oxides add particulate mass to containment pool, which may enhance coagulation.
T1-2	RCS coolant conditions at break	The RCS coolant chemistry varies over the fuel cycle. B concentrations $\approx$ 2000 -4000 ppm (beginning) to 50 ppm (end of fuel cycle) B/Li ratio $\approx$ 100. The 2-phase jet is at temperatures of $\approx$ 315C to 120C upon cooling	Initial reactor water chemistry spewing out of the break and forming the containment pool will have variable B concentration. The Li/B ratio is maintained approximately constant.
T1-3	pH variability	pH @25C at beginning of fuel cycle is acidic (4) while closer to neutral (7) at end of cycle.	Similar implications to issue 2; variable reactor and initial containment pool chemical environment.
T1-4	Localized B concentration in jet	Droplet evaporation within the high temperature jet emanating from the pipe break forms at highly concentrated B aerosol solution. As boric acid concentrates at high temperatures (250C - 300C), a polymerization appears to occur that releases protons and causes the system to become very acidic.	1. Corrosion of impacting surfaces: enhanced erosion & corrosion vs. formation of a protective lacquer to limit erosion. 2. May limit B concentration in containment pool chemistry
T1-5	RCS fluid creates "oxidizing environment"	Coolant is released as a super heated spray once break occurs. Hydrogen (25 – 35 cc/kg of H_2 injected as gas) transfers to the gas phase.	1. H_2 evaporation creates a more oxidizing environment in the water phase by raising the redox potential. 2. The partial pressure of H_2 is lowered because of the large volume of the gas phase (containment volume) compared with the water volume being released. This also increases the redox potential.

ID	Topic	Description	Effects
T1-6	Jet impingement	Water jet & fine debris within jet may impact surfaces and chip coatings	Initiates metallic pitting, corrosion and ablation of other materials (concrete, insulation)
T1-7	Break proximity to organic sources	RCP oil storage tank is made with ≈ ½" carbon steel with epoxy coating (250 gallon per pump, but tank may just contain leakage).	Tank failure timing: earlier in post-LOCA (jet impingement) vs. later in post-LOCA (corrosion) Organic sources to containment pool would alter chemistry (i.e., complexation, etc.)
T1-8	Break proximity to secondary systems	Break near cooling systems could result in failure of lines containing solvents (i.e., Freon) into containment environment	1. Radiolytic decomposition could create chlorides and fluorides which may impact corrosion and containment pool chemistry. 2. Possible effects on pump performance
T1-9	Debris mix particle/fiber ratio	Breaks in different locations will create different debris characteristics: total mass, mixture constituents, compositions	1. Could alter the containment pool chemistry. 2. Could affect debris bed formation on the sump screen and increase variability of chemical product capture efficiency.
T1-10	Hydrogen peroxide effects	Hydrogen peroxide will behave much the same as H_2O and most of it will remain in the liquid phase. Additional hydrogen peroxide evaporates into containment vapor space and may cause additional corrosion. It condenses on cooler surfaces and can accelerate corrosion there	1. Could affect how much H_2O_2 is in the containment pool: Evaporation & condensation could sequester form pool droplets could transport back to pool. 2. H_2O_2 alters containment pool chemistry and potentially causes changes in oxidation states or chemical forms in sump water..
T1-11	Nuclei Formation	Rapid subcooling and evaporation of water from effluent of broken pipe droplets might lead to precipitation of LiOH and other species. Concentrations increase and temperature decreases rapidly	1. Initial precipitate may be formed from species existing with in the RCS coolant. 2. Nuclei could form which could enhance precipitation of other species within the containment pool at a later time.

Table E-2: Apps T1 Ranking Summary

Issue Number	Importance (H, M, L)	Rationale	Knowledge (K, PK, UK)	Rationale
T1-1	L	Particles would be dense and tend to settle out.	PK	Extent of release cannot be accurately predicted.
T1-2	L	Effect is transient. Will be compensated by addition of cooling water.	K	Plant operating parameters well known.
T1-3	L	Effect is transient. Will be compensated by addition of cooling water and pH buffers.	K	Plant operating parameters well known.
T1-4	L/M	Effect is transient. Depends on B concentration. Will be compensated by addition of cooling water and pH buffers.	PK	Actual behavior during LOCA not quantified.
T1-5	L	Quantity of H_2 is small.	K	Plant operating parameters well known. Predictable outcome.
T1-6	H	Principal mechanism for generating reactive particulates.	PK	Actual extent of damage would vary.
T1-7	M/H	Release of oil could significantly affect corrosion and debris flotation.	UK	Likelihood of release is speculative.
T1-8	L	Freon would partition into the containment-building atmosphere and be relatively non-reactive.	UK	Radiolysis in gas phase probably low.
T1-9	H	Debris composition and amount would strongly affect screen blockage and formation of secondary precipitates.	PK	Potential damage scenario can be postulated. Depends on plant design.

T1-10	L	H_2O_2 reactive, but quantity present may be small. H_2O_2 may induce passivation of metals. Time for reaction to occur is limited.	PK	H_2O_2 concentration can be predicted, but corrosion rates of metals exposed to H_2O_2 would need to be evaluated.
T1-11	L	Particles may re-dissolve. Effect probably minor over short time frame.	UK	Require evaluation to determine whether the phenomenon is significant.

Table E-3: Chen T1 Ranking Summary

Issue Number	Importance (H, M, L)	Rationale	Knowledge (K, PK, UK)	Rationale
T1-1	M		PK	
T1-2	M		K	
T1-3	M	pH will affect the rate of corrosion of metals	K	
T1-4	M		PK	
T1-5	L		PK	
T1-6	L	The impact areas may not be that significant	PK	
T1-7	L		PK	
T1-8	M		PK	

T1-9	H	Debris mix definitely affect the amount (thickness) and characteristics (tightness, compressibility,..) of the cake on the sump screen	K	
T1-10	L		PK	
T1-11	L		PK	

Table E-4: Delegard T1 Ranking Summary

Issue Number	Importance (H, M, L)	Rationale	Knowledge (K, PK, UK)	Rationale
T1-1	H	Not considered in ICET. Significant potential solids source also includes fuel surfaces.	PK	Crud releases observed during reactor shut-down and spent fuel storage.
T1-2	H	Boron concentration wide variability and high temperature not considered in ICET but B has large affect on solids corrosion.	PK	Knowledge based on ICET results only for set B concentration and temperature.
T1-3	H	Boron concentration wide variability and high temperature not considered in ICET but B has large affect on solids corrosion.	PK	Knowledge based on ICET results only for set B concentration and temperature.
T1-4	M	Impact of localized higher B concentration limited by small affected areas.	PK	Studies of higher B concentrations not performed.
T1-5	L	Dissolved H_2 concentration ~25-35 cm^3/kg water at 0°C and much lower at temperature so reducing conditions from H_2 unlikely.	PK	Real issue is overall Eh of system and this has not been measured under radiolytic conditions.
T1-6	H	This phenomenon should add large quantities of debris to reactor coolant.	PK	Testing needed (it would seem) to judge effects of steam jetting on metal, paint, insulation, and concrete.
T1-7	M	Effects of oil not studied; oil may accrete small solids and drift downstream to span matted fibers at sump screen.	PK	Effects of oil to accrete finely particulate solids under this system not known but example exists in oil spills from oil tankers onto beaches.
T1-8	L	Freons are chemically rugged and, even if released, unlikely to react significantly.	PK	Radiolysis may have impact on Freon stability but access of Freon to high fields may be limited.

T1-9	H	Though important, this question likely addressed in sufficient detail in ICET.	K	ICET gives good starting point.
T1-10	L	H_2O_2 is not very volatile and is unstable to disproportionation to form O_2 and H_2O, especially in the presence of iron.	K	Effects of H_2O_2 in the coolant system should be studied but the efficient transport of H_2O_2 out of coolant unlikely to have much effect.
T1-11	M	This is similar to T1-4, only for LiOH and other species.	PK	Studies of higher solute concentrations not performed.

Table E-5: Litman T1 Ranking Summary

Issue Number	Importance (H, M, L)	Rationale	Knowledge (K, PK, UK)	Rationale
T1-1	H	Thermo-hydraulic transient and formation of hydrogen peroxide absent hydrogen overpressure will cause significant CRUD release.	K	This is similar to an unplanned reactor trip event. The exception is water temperature during a reactor trip remains elevated and hydrogen control is maintained minimizing peroxide formation.
T1-2	H	At beginning of cycle with High [B], pH of RCS as it impinges will be fairly acidic and hot (high temperature). This will cause significantly greater corrosion and leaching from impacted materials.	K	During the early phase of the break reactor coolant without buffering agent is sprayed directly onto ZOI where the greatest surface area contact will occur on insulating materials (principally responsible for the precipitates). The high temperature will increase leaching and solubility of ionic materials.
T1-3	H	Same as above.	K	Same as above. However, this changes with time. Final pH is significantly different than the initial and depending upon the media impacted will have different effects.
T1-4	L	Although this is technically correct, it will have a low overall impact on debris generation and reaction with the debris.	K	Physical properties of boric acid are well established.

T1-5	M	Hydrogen solubility decreases dramatically due to pressure drop. Within reactor containment if hydrogen level gets too high recombiners go on to eliminate explosive mixture. With no hydrogen in the liquid phase and peroxide production in the core, redox potential goes positive by 100+ millivolts.	K	The formation of peroxide is a well-established phenomenon in the SFP of nuclear power plants. Hydrogen loss is a physical property.
T1-6	L	Reactor containment buildings are 'crowded'. The available coating surface that would impacted by the initial jet spray is small compared to the total surface area of painted surfaces in the containment.	PK	This is a plant specific parameter. It would be more significant for plants that had variable insulation types that were in close proximity to the break.
T1-7	L	This is a possbility depending on where the break occurs and the proximity to these material sources. The tanks themselves are carbon steel and coated with epoxy.	PK	This is a plant specific parameter.
T1-8	L	The projected break vicinity is usually distant from component cooling systems or secondary systems	PK	This is a plant specific parameter. It depends on the type of materials in closed cooling systems in containment.
T1-9	M	This is dependent upon the type and amount of insulation adjacent to the ZOI of the break	PK	This is a plant specific parameter.
T1-10	M	The presence of hydrogen peroxide on non-submerged surfaces will enhance corrosion rates. However this would be most significant during the first four hours when spray is washing down surfaces. After that it will be of much less importance.	K	The effect of peroxide on metallic components corrosion can be estimated from corrosion rates.

T1-11	L	The effects of LiOH at ambient temperatures will be minimal especially since the RWST boron will rapidly put the break chemistry into the acidic range.	K	LiOH is a weak base at room temperature. The RCS $pH_{308\,C}$(boron 1200 ppm and Li 3.5 ppm) is ~7.1 above a $pH_{neutral}$ of 6.8. The same solution is pH of ~5.1 below $pH_{neutral}$ of 7.0

Table E-6: MacDonald T1 Ranking Summary

Issue Number	Importance (H, M, L)	Rationale	Knowledge (K, PK, UK)	Rationale
T1-1	H	Radiolysis can considerably modify the chemistry of the water in the containment pool by raising the redox potential. A higher redox potential will result in the oxidation of lower valency cations (e.g., Fe^{2+}) to higher valency cations (Fe^{3+}), which subsequently hydrolyze to precipitate solids (e.g., FeOOH) that may clog the pump screens. The hydrolysis of Fe^{2+} will produce $Fe(OH)_2$, which in turn may undergo the Schikorr reaction to form magnetite (Fe_3O_4) and hydrogen. The latter will contribute to the total hydrogen inventory and will impact the redox potential. Some scoping calculations of the radiolysis of water and stainless steel corrosion potentials, performed during the present study for illustrative purposes, are summarized in the author's report.	K	Radiolysis and mixed potential models are available for estimating the redox potential in the containment pool, data and computer codes are available for modeling the hydrolysis of many common ions (e.g., Fe^{3+}, Ni^{2+}, Cr^{3+}, Cu^{2+}, Zn^{2+}, and Al^{3+}) and the formation of the corresponding hydroxides, oxyhydroxides, and oxides.

E-13

T1-2	H	The Li/B ratio in the coolant, together with the NaOH, TSP, and STB buffers stored on the containment floor, will determine, to a significant extent, the acid/base properties (pH) of the coolant in the containment pool. Thus knowledge of the Li/B ratio at the time of the LOCA is vital for modeling the chemistry of the water in the containment pool, since it describes the "initial state". The processes that occur as the coolant expands adiabatically upon a pipe break are also of great significance.	PK	The Li/B ratio is known precisely at any point in the fuel cycle, as are the physico-chemical properties of the coolant. However, the adiabatic expansion of the coolant, upon occurrence of a LOCA, accompanied by evaporation of water and hydrogen transfer from the coolant droplets, is expected to significantly modify the chemical (e.g., pH) and electrochemical (redox potential) properties of the coolant that eventually collects in the containment pool. The physico electrochemical processes involved are poorly defined and understood at the present time, even at the fundamental level.
T1-3	H	See T1-2	PK	See T1-2.
T1-4	H	Concentrated boric acid is known to be corrosive to ferrous alloys (e.g., carbon steel), as witnessed by the Davis Besse reactor head degradation phenomenon, and is postulated to be corrosive toward many other materials that are susceptible to highly acidic conditions. Thus, high temperature pH measurements on concentrated boric acid detected the occurrence of a polymerization process that released protons (H^+).	PK	While the production of highly acidic conditions in concentrated boric acid have been detected in laboratory experiments, so little is known of this phenomenon that we do not have the capability of modeling the damage that concentrated boric acid that is formed in a LOCA jet may cause.

T1-5	H	The chemistry of the coolant expanding from a LOCA break is determined to a significant extent by the redox potential of the water (for example, the redox potential determines the valency of metal corrosion products). In turn, the redox potential reflects a balance between the oxidizing (e.g., O_2, H_2O_2, Fe^{3+}) and reducing (H_2, Fe^{2+}) species in the solution. Thus, as H_2 is transferred from the solution to the gas phase, and noting that, because H_2 has a lower molecular weight than do O_2 and H_2O_2, the droplets in a LOCA stream will gradually become more oxidizing in nature. In consequence, Fe^{2+} will be oxidized to Fe^{3+} and FeOOH will be precipitated.	PK	Theories have been developed for the transfer of gasses from the dissolved state in a solution to the gas phase, but it is not known whether these models can be used to predict gas transfer under the "shock" conditions of a LOCA or whether the requisite data are available for the model parameters to allow effective prediction.
T1-6	H	The erosion of materials by the LOCA jet is deemed to be a significant source of debris in the containment pool immediately after a LOCA. The rate of damage accumulation (debris production) is a sensitive function of many variables, including jet velocity and composition, impact angle, distance from the jet source and point of impact, and so forth.	K	The erosion-corrosion of materials by single-phase (liquid, gas) and by two-phase (liquid + gas, liquid + solid, and gas + solid) jets has been extensively studied, at both the theoretical and practical levels. Models are available (e.g., the Keller equation), but the applicability of these models to a LOCA scenario has yet to be determined.
T1-7	H	See T1-6.	K	See T1-6.

T1-8	L	The amount of radiation-induced damage at the prevailing γ-dose rates within the first 30 seconds of a LOCA are expected to cause minimal damage to chloride- and fluoride-containing polymers, so that the release of Cl⁻ and F⁻ from radiolytic processes is likewise expected to be minimal, compared with other processes (e.g., thermo hydrolysis).	PK	The rates of radiation-induced damage to many plastics and polymers have been characterized over wide ranges of conditions. These data are readily available in the public domain. However, there is the possibility of synergistic interaction between radiation-induced damage and thermo hydrolysis processes that have not been significantly explored.
T1-9	M	Small variations in the contents of the containment pool are probably not very significant at this level of analysis. It is expected that all LOCA events will produce silicates and alumino-silicates (eroded concrete), fiber glass, paint and polymer debris, metal corrosion products, etc).	PK	Depends on the location of the break, which is not predictable at the current time, other than there being an increased l kelihood that it will occur at a highly stressed point in the main coolant line.

T1-10	M	Hydrogen peroxide is a strong oxidizing agent and hence is capable of inducing severe corrosion on metals, such as Al, and on alloys, such as the stainless steels. H_2O_2 may be transferred to the containment volume via spray and evaporation. Initial rankings changed due to following email, rankings indicated separately: Please find attached a table with my re-assessments of some originally "high" ratings of issues. On examining these issues again, it was evident that I did not fully recognize the impact of: 1. Time of exposure after the initial LOCA. 2. The amount of the specific material in the containment (e.g., zinc). My re-assessments also reflect the results of continuing radiolysis modeling, albeit with less-than-optimal codes that have been jerry-rigged from our reactor modeling work. Nevertheless, if the fields are as high as Bob indicates, the eventual production of nitric acid, in addition to the elevation of the redox potential, becomes a critical issue.	K	The impact that H_2O_2 can have on the corrosion and degradation of materials has been studied extensively and many data and much information are available in handbooks. Furthermore, the theoretical basis for predicting corrosion rates in environments containing hydrogen peroxide is highly developed in the form of the Mixed Potential Model (MPM) and the Point Defect Model (PDM).

T1-11	L	The kinetics of precipitation of solids from super-saturated solutions are highly dependent upon the type and volume density of nuclei. Use of this fact is made in numerous industrial processes by seeding a solution to induce precipitation and hence to separate a wanted or unwanted product from the system. Thus, the inventory of solid products in the containment pool within the first 30s of a LOCA will depend sensitively on the concentration and properties of the nuclei. In the absence of effective nuclei, as super-saturated solution may remain metastable over extended periods (minutes to months). Initial rankings changed due to following email, rankings indicated separately: Please find attached a table with my re-assessments of some originally "high" ratings of issues. On examining these issues again, it was evident that I did not fully recognize the impact of: 1. Time of exposure after the initial LOCA. 2. The amount of the specific material in the containment (e.g., zinc). My re-assessments also reflect the results of continuing radiolysis modeling, albeit with less-than-optimal codes that have been jerry-rigged from our reactor modeling work. Nevertheless, if the fields are as high as Bob indicates, the eventual production of nitric acid, in addition to the elevation of the redox potential, becomes a critical issue.	PK	An enormous literature exists on the theory and practice of the nucleation and solid precipitation in multi-component systems, but mostly for "simple" systems (e.g., salts). How well and how easily these theories and experimental databases can be applied to LOCA scenarios remains to be determined.

E-19

APPENDIX F

PIRT Panelist Evaluations of T2 Phenomena: ECCS Injection from 30s into LOCA until ECCS Recirculation Begins

Table F-1: ECCS Injection (30 seconds to t_{recirc})[a,b,c]

Issue Number	Phenomena	Description	Implications
T2-1	Hydrogen sources within containment	H_2 concentrations in vapor and containment pool include the RCS inventory, Schikorr reaction, corrosion of metallic materials	Containment pool redox potential is a function of dissolved hydrogen which is established by the listed reactions. RCS evaporation may lead to equilibration with liquid, decreasing redox potential
T2-2	ECCS injection of Boron	After pipe break, B ($\approx$ 2800 ppm) is injected into RCS to cool reactor	Provides large B source which may affect chemical reaction products in containment pool. Specifically, the B source will serve as a pH buffer.
T2-3	Containment spray corrosion	Containment sprays for buffering creates wetting & surface films on material surfaces with water containing B, LiOH, H_2	Containment spray (up to 4 hours) could enhance corrosion formation on unsubmerged metallic species which could contribute to containment pool concentrations. Spray pH: NaOH = 8.5 – 10.5 TSP = 4.5 - 5 STB = 4.5 – 5
T2-4	NaOH pH control	NaOH injected through containment sprays.	Alkaline containment pool environment (pH = 8.5 – 10.5)
T2-5	STB pH control	Released with ice in containment during a LOCA	Alkaline containment pool environment. Initially pH is high 9 initially and then decreases with mixing (pH = 8 – 9)
T2-6	TSP pH control	Contained in basket on containment floor that dissolve	Slightly alkaline containment pool environment. Initially pH is high (pH = 11) until complete mixing within about 4 hours (pH = 7 – 8)
T2-7	Containment spray transport	Sprays wash latent debris, corrosion products, insulation materials, & coating debris into containment pool	Affect on containment debris sources (types, amounts, compositions) and contributions to the chemical sump pool environment.
T2-8	Containment spray CO_2 scavenging	Containment sprays cause CO_2 absorption within containment pool and carbonate formation.	Affect on containment pool concentrations of CO_2
T2-9	Debris dissolution begins	Debris dissolution begins. Initial expected products include cal-sil, cement dust, organic fiberglass binders, epoxy & alkyd coatings, uncoated concrete	Indicate potential important contributors (if any) to chemical containment pool environment during this time frame. Dissolution of other products will occur over longer time frames.

T2-10	Carbonate concentration	Carbonates in solution are provided by atmospheric CO_2, and concrete dust at this point	1. Create nucleation sites for species precipitation 2. scavenge dissolved Ca to impede formation of other compounds to form calcium carbonate
T2-11	Containment pool mixing	High velocity, turbulent water flow exists underneath the pipe break location, remainder of the pool relatively quiescent	Mixing of aqueous/gas phases at the break location is enhanced; could aid CO_2 absorption; buffering agent dissolution/mixing, liquid/vapor exchange of species.
T2-12	Boric acid corrosion of exposed concrete	Concentrated boric acid solutions in contact with exposed concrete	1. Calcium boro-silicates may precipitate 2. pH decrease in containment pool possible
T2-13	Fe, Ni radiological reaction	Initial Fe, Ni oxide breaks off due to TH transient. Reduced Ni, Fe is dissolved in RCS, combines with air, oxidizes and forms hematite, maghemite, and magnetite. . Kilorad/hour and higher rad. dose rates arise throughout containment due to dispersal of Fe and Ni activation products.	1. Solid particulate is added to containment pool, another debris source 2. High pH will increase the reaction. 3. Reaction continues until RCS piping cools down (1 day or more). 4. Radiolysis products generated throughout post-LOCA cooling system
T2-14	Hydrolysis	Nickel oxide becomes catalyst for producing H_2 from radiolysis	Affects redox potential
T2-15	Organic Complexation	Organic acids become absorbed on surfaces of solids & inh bit solid species growth; example: aliphatic acid	Effectively enhances solubility limits. Solid species may precipitate, but remain relatively small in size (nano-scale). They do not agglomerate or grow to macroscopic sizes.
T2-16	Organic Sequestration	Organic electron-rich atoms combine with soluble metal compounds.	Sequesters metal ions so that solid species precipitation is delayed or does not occur.
T2-17	Auxiliary Component Cooling Line Failure	LOCA causes attached or nearby component cooling line failure which could release a number of chemicals into the containment pool (chromates, molybdates (6000 – 1000 ppm in storage tank), nitrites, tolytriazole (25 ppm in storage tank), benzotriazole, hydrazine	1. The release of these types of chemicals could increase the ionic strength of the containment environment which may decrease gel stability. 2. Chemicals may accelerate corrosion.

T2-18	Polymerization	Precursor to precipitation & agglomeration; metals/oxygen bonds, ripen to form covalent bonds, growth until they qualify as ~nanometer particles size: Si, Al, Fe, Boric acid are candidates	May be necessary to form large enough particles to result in tang ble effects on ECCS performance
T2-19	Co-precipitation	Method of precipitation/separation Examples: Ni/Fe/Cr, Al/Si/B; Co/Fe systems; Precipitation of one species leads to precipitation of other species (below solubility limit); Radioactive elements precipitate (at low concentrations), activation products: ex. Strontium, Ni, Silver, Fe, Co, Zr	More solids species form which could lead to greater concentration of chemical products at the sump screen or downstream.
T2-20	Radiolytic environment	Radiolysis reaction creates an oxidizing environment and can change solution's electric potential; H_2, O_2, and H_2O_2 balance most important for determining redox potential	Could affect chemistry in containment pool which affects species which form (e.g., Hanford tank: mixture of different components @ different phases)
T2-21	Inorganic Agglomeration	Formation of larger clumps of smaller particulates: Depends on PZC; ionic strength (the higher the strength the smaller the distance for agglomeration)	1. May be necessary to form large enough particles to result in tang ble effects on ECCS performance 2. Can occur quickly if conditions are right. 3. Existence of organic species can increase or decrease likelihood (See T2-24)
T2-22	Galvanic Effects	Electrical contact between metals results in increased corrosion of less noble metals: example: magnetite precipitation on Aluminum	1. Increases in the corrosion source term for less noble (anodic) metal. 2. Decreases in corrosion source term for more noble (cathodic) metal.
T2-23	Deposition & Settling	Poss bility that chemical products formed during this time period either settle within containment pools or are deposited on other surfaces	Potentially remove substantial quantities of material from transporting to the containment sump screen: indicate expected species that are most likely to settle, if poss ble.
T2-24	Organic Agglomeration	Formation of larger clumps of smaller inorganic particulates nucleating around organic acids or oil (such as soap coagulates dirt particles)	Coagulated particles can collect on sump screen, decreasing flow, or collect in other places to decrease the loading on the sump screen

[a] Massive fuel damage is not considered during this time period because the objective is to prevent this through successful operation of the ECCS system. Radiological and other contributions to the chemical environment from minor amounts of fuel damage are expected to be insignificant compared to crud release and other phenomena listed in these tables.
[b] Sump pool temperatures can be as high as 120C initially. After 30 minutes, between 60 – 90C
[c] Phenomena listed in earlier time periods (i.e., T1) may continue to be active or important during this time.

Table F-2: Apps T2 Ranking Summary

Issue Number	Importance (H, M, L)	Rationale	Knowledge (K, PK, UK)	Rationale
T2-1	L	Quantity of H_2 is small. Partitioning into containment building atmosphere would render it largely inert.	PK	Plant operating parameters well known. Behavior of hydrogen generally predictable.
T2-2	H	B would have a significant effect on coolant chemistry.	K	ICET series provide useful data on chemical interactions with B.
T2-3	L	Effect is transient.	K	ICET series indicates that unsubmerged corrosion is generally minor.
T2-4	M	NaOH would have a significant effect on coolant chemistry, but time period is too short for substantial reaction to occur.	K	ICET series provide useful data on chemical interactions due to presence of NaOH
T2-5	M	STB would have a significant effect on coolant chemistry, but time period is too short for substantial reaction to occur.	K	ICET series provide useful data on chemical interactions due to presence of STB
T2-6	M	TSP would have a significant effect on coolant chemistry, but time period is too short for substantial reaction to occur.	K	ICET series provide useful data on chemical interactions due to presence of TSP
T2-7	H	Quantity and make-up would strongly affect sump screen blockage.	PK	Consequences are difficult to predict following a LOCA. Evaluations need to be made on a case-by-case basis.
T2-8	L/M	Quantity of CO_2 is low, but could induce formation of $CaCO_3$ seeds for nucleation and growth of other phases.	PK	Amount of CO_2 that could be absorbed is well known, but the consequent effect on nucleation and precipitation is less clear.

T2-9	M	Debris dissolution is instrumental in the formation of secondary precipitates. However, extent of reaction during the initial phase would be small	K	ICET program has evaluated debris corrosion rates.
T2-10	L/M	Quantity of CO_2 is low, but could induce formation of 3 seeds for nucleation and growth of other phases.	PK	Amount of CO_2 that could be absorbed is well known, but the consequent effect on nucleation and precipitation is less clear.
T2-11	H	Debris settling would limit sump screen blockage.	UK/PK	Consequences are difficult to predict following a LOCA. Evaluations need to be made on a case-by-case basis.
T2-12	L	Secondary calcium borosilicates may actually protect concrete from further damage. However, time period is too short for significant reaction to occur.	UK	Effect would require study.
T2-13	L	Quantities are small. Amounts of precipitates are small.	PK	Extent and magnitude of the effect would require further study.
T2-14	L	Effect of H_2 would be small. H_2 would partition into the containment-building atmosphere, and would be relatively inert. Minimal effect on redox potential.	UK	Effect would require evaluation under LOCA conditions.
T2-15	L/M	Amount of organic acid would be minor, but possibly significant effect on surface reactivity.	PK	Consequences would require further study.
T2-16	L	Amount of organic acid would be minor, and overall effect would be small	UK	If the identity and concentrations of organic acids are known, then the effect on metal solubility could be calculated.
T2-17	L/M	Quantities released would be small in comparison with circulating volume.	PK	The impact of added chemicals would require evaluation.

T2-18	H	Precursors to nucleation and precipitation.	PK	ICET series and associated head loss studies provide some information.
T2-19	L/M	Co-precipitation of radionuclides could lead to their concentration in precipitates at the sump screen. However, radionuclide concentration is likely to be low.	PK	Effect of co-precipitation can be estimated in some systems. Experimental studies would be required to determine the impact of co-precipitation in most cases.
T2-20	L	Corrosion rates of Al could be enhanced by radiation induced elevated oxidation state. However, the environment is already oxidizing.	PK	Effect would require quantification for specific LOCA conditions.
T2-21	M/H	Could affect sump screen blockage.	PK	ICET series and associated head loss studies provide some information.
T2-22	L/M	Could be significant depending on the design of the PWR. Short-term effects would be small.	PK	Phenomenon needs evaluation in the context of a LOCA.
T2-23	M/H	Settling would affect sump-screen blockage.	UK	Effect is specific to the plant design.
T2-24	L/M	Effect depends on the type and magnitude of organics released.	UK	Depends on the conditions following a LOCA, and plant design.

Table F-3: Chen T2 Ranking Summary

Issue Number	Importance (H, M, L)	Rationale	Knowledge (K, PK, UK)	Rationale
T2-1	L		PK	
T2-2	L		PK	
T2-3	M		K	
T2-4	L	Too short a time	K	
T2-5	L	Too short a time	K	
T2-6	L	Too short a time	K	
T2-7	H	This affects the amount of debris to the sump screen	PK	
T2-8	No comment		No comment	

T2-9	L		PK	
T2-10	L		PK	
T2-11	L		PK	
T2-12	M		PK	
T2-13	M		PK	
T2-14	L		PK	
T2-15	L		PK	
T2-16	M		PK	
T2-17	L		PK	
T2-18	M		PK	

T2-19	M		PK	
T2-20	L		PK	
T2-21	L		PK	
T2-22	No comment		No comment	
T2-23	M		PK	
T2-24	H		PK	

Table F-4: Delegard T2 Ranking Summary

Issue Number	Importance (H, M, L)	Rationale	Knowledge (K, PK, UK)	Rationale
T2-1	L	H_2 concentration in solution low (see T1-5) and H_2, in the absence of a catalyst such as platinum metal, is not a very effective reductant.	PK	Real issue is overall Eh of system and this has not been measured under radiolytic conditions.
T2-2	H	Boron has demonstrated large effect in ICET and other testing.	K	Boron has demonstrated large effect in ICET and other testing.
T2-3	L	H_2 was already considered in T2-1; other constituents shown to have low effects in ICET.	K	Already studied by ICET and other testing.
T2-4	H	The NaOH is large contr butor to pH balance and solute loading.	K	Studied by ICET and other testing.
T2-5	M	The sodium tetraborate is a large contributor to solutes. I downgraded Delegard from high to medium based on email stating that this issue is secondary in light of the others and/or because it has been considered in other time frames.	K	Studied by ICET and other testing.
T2-6	H	The trisodium phosphate is a large contributor to solutes.	K	Studied by ICET and other testing.
T2-7	M	Expect that the loose dust and other solids will be relatively small additions to the solids loosened by initial spray release.	PK	Dust and loose debris loading in the reactor containment building not estimated.

T2-8	M	For CO_2 absorption to have a significant effect, it must find a Ca^{2+} source to react with. Best source is Cal-Sil. Net effect of reacting CO_2 with Cal-Sil is conversion of $Ca(OH)_2$ solids from dissolution/precipitation reaction of Cal-Sil in water with dissolved CO_2 to form $CaCO_3$, effectively trading one finely particulate solid for another.	K	CO_2 budget in the largest reactor containment building is ~170 kg and about 50 kg in most reactor containments to increase the solids loading ~100 kg and 30 kg, respectively, by converting $Ca(OH)_2$ + $CO_2 \rightarrow CaCO_3 + H_2O$. This is likely a small contributor to the total solids released and generated in a LOCA.
T2-9	H	Demonstrated by ICET and other testing.	K	Demonstrated by ICET and other testing.
T2-10	M	These effects likely are moderate compared with other phenomena.	K	These effects likely are already demonstrated by ICET and other tests.
T2-11	L	Effect of turbulence is better mixing. This effect is also achieved by the pump and thus is already accounted.	K	Effects of mixing built into ICET testing in this time frame (note, however, that ICET did not simulate the initial, Table 1, high energy jet impingement).
T2-12	M	Effects of borates on concrete (and other surfaces such as fiberglass and aluminum) important.	PK	Lab testing began showing influence of boron.
T2-13	M	Though title of problem mentions radiolysis, text describes chemical phenomena. Concentrations of dissolved Ni, Fe, and other metals in coolant are low and their precipitation will add little to the total solids budget.	PK	Some phenomenological information exists from refueling observations.
T2-14	L	The effect of H_2, as mentioned in T1-5 and T2-1, is likely to be small despite the additional effects of H radical combination of Ni or other surfaces.	PK	Real issue is overall Eh of system and this has not been measured under radiolytic conditions.

T2-15	H	The organic acids, in addition to absorbing on surfaces and engaging in complexation (as descrbed by T2-16), also might act as surfactants to coagulate or agglomerate particles. Additional rationale in email response: (Organic Complexation; which is more aptly titled Organic Coating or Surfactant): I still believe the effects of organic decomposition products that deposit onto solid particles could be significant on solids growth, both to increase growth or agglomeration or inhibit further growth or agglomeration. The conditions associated with this phenomenon have not yet been investigated in sufficient detail because conditions leading to release of the organics (high-temperature water and radiolysis) have not yet been studied.	PK	The sources of organic acids (and other organic species) and their effects have not been considered in testing to-date but can have large effects on the chemical and mechanical properties of the largely inorganic coolant system behavior.
T2-16	M	See T2-15. I downgraded Delegard from high to medium based on email stating that this issue is secondary in light of the others and/or because it has been considered in other time frames.	PK	See T2-15.
T2-17	M	The components in the cooling lines may have a significant effect but their concentrations are small.	PK	Testing needed to see if the components have an effect. The effects on chromate and molybdate on aluminum corrosion could be large.
T2-18	M	Demonstrated by ICET and other testing.	K	Demonstrated by ICET and other testing.

T2-19	M	Demonstrated by ICET and other testing.	K	Demonstrated by ICET and other testing.
T2-20	M	The effects of radiolysis on the inorganic constituents likely are small but likely are large for the organic materials (e.g., paint, plastics, oils).	UK	Testing needed here.
T2-21	M	Demonstrated by ICET and other testing.	K	Demonstrated by ICET and other testing.
T2-22	M	ICET specifically avoided galvanic conditions.	UK	Testing of galvanically coupled metals needed under post-LOCA conditions.
T2-23	L	This actually could be a large effect but there does not seem to be a credible way to know, for the various reactor configurations and LOCA conditions, how to model the flow routing and catchments and weirs.	UK	This potentially important phenomenon could afford a way to minimize solids transport to the sump screens and invites study of methods, outside of the scope of the chemical effects testing, to exploit this (e.g., "carpeting" the floor leading to the sump screen to snag solids from the coolant).
T2-24	H	See T2-15. This mechanism has a large potential to grow solids sizes and block sump screen passages.	UK	Testing is needed here for both raw organic materials (such as lubricating oils) and thermally and radiolytically degraded organic materials (e.g., paints, plastics).

Table F-5: Litman T2 Ranking Summary

Issue Number	Importance (H, M, L)	Rationale	Knowledge (K, PK, UK)	Rationale
T2-1	L	The rate of hydrogen production due to corrosion will decrease with time. Containment buildings have hydrogen recombiners that eliminate explosive potential. Most are set to start at 2% hydrogen in the containment atmosphere. This will lead to negligible dissolved hydrogen to affect ORP.	K	Dissolved hydrogen can be calculated from containment pressure, % H_2 and Henry's Law
T2-2	M	The 2800 ppm B will yield a pH of 4.5-5.5 depending upon temperature. This will cause hot, acidic corrosion of metal components.	PK	This is a plant specific parameter based on the components in the ZOI.
T2-3	M	This will differ based on the containment spray systems. The TSP based spray will go for at least 30 minutes with no buffering thus pH will be 4.5-5.5; this will be significant. In ice condensers and NaOH plants the spray will be buffered.	PK	This is a plant specific parameter based on the type of buffer system used.
T2-4	H	For NaOH plants the spray will cause accelerated corrosion of unsubmerged materials. This means high dissolved metals content which subsequently may precipitate upon cooling.	K	The ICET program has shown that the highest TDS will result from the NaOH spray program.
T2-5	L	The STB plants will have a spray that is buffered at a more neutral pH.	K	ICET program has shown that the least amount of TDS and precipitated materials with this spray buffer.

T2-6	H	TSP containment buffer systems provide the highest degree of precipitate formation.	K	The phosphates of the major metal ion components of insulation and corrosion products form insoluble precipitates. This adds to the mass of material that can potentially be transported to containment screens.
T2-7	L	The amount of 'loose' debris in containment that could be washed into the sump is minimal. NRC enforces the amount of miscellaneous material that can be stored in containment, and that material in containment must be anchored.	K	Each plant will have a good idea of what potentially will be removed as debris.
T2-8	M	The carbon dioxide concentration of air provides an additional 69 pounds of insoluble material just from the carbonate mass. This does not include the counter ion and any co-precipitated materials.	K	The maximum mass can be calculated based on the type of insulation.
T2-9	M	The greater the mass of leached materials that can contribute to insoluble mass the more concern. Cal-Sil and TSP are the worst combination.	K	ICET program confirms that the calcium phosphate amorphous material provides the greatest impediment to flow through a filtering medium.
T2-10	M	In the alkaline environment of all three buffers carbonates will precipitate. It appears to be most significant at pH 10 with NaOH.	K	The higher the pH the more effective and rapid will be the carbon dioxide scavenge.
T2-11	M	TSP will have the most significant effect since the phosphate concentration will be relatively high at sump until the recirculation phase is initiated. Significant precipitation could occur with Cal-Sil if enough damage is done to insulation.	K	The results of the ICET program confirm this.

T2-12	L	Few areas of containment buildings are bare. They also have minimal surface area. Boric acid will neutralize the leached materials from bare concrete (which will be basic).	K	pH of exposed concrete with demineralized water is pH 9-10.5. Boric acid will buffer this.	
T2-13	M	The presence of the insoluble CRUD at this phase will have significant effects. It will help to agglomerate other insoluble materials due to the density of this material.	K	CRUD layer in PWRs is very susceptible to dislodging following thermo-hydraulic and pH transient of a reactor trip.	
T2-14	L	The production of hydrogen would be very slow. The hydrogen produced would not reach concentrations above 3-4% due to hydrogen recombiners in containment.	K	Reaction rate will be based on water in contact with the NiO. This is dense material and will precipitate. Plant design prevents significant hydrogen build up.	
T2-15	L	The likelihood of a large organic contribution to the containment volume is small regardless of the type of insulation or buffering agent. The radiolysis of the organics could yield organic acids (complexing agents) or carbon dioxide (precipitating agents as carbonates)	UK	No quantitative measure can be ascribed to this since it depends upon the plant, the plant design, the particular organic materials that would be present and their susceptibility ot total or partial radiolysis.	
T2-16	L	Same as T2-15	UK	Same as T2-15	
T2-17	M	PWRs have significantly different mixtures of chemicals in the component cooling water systems, and there are significant volumes of these systems in the containment building. At this stage of the event, radiolysis would not have taken place as these compounds would not have recirculated through the core.	UK	No significant effects are expected in the short term due to the short contact time which would be found in the sump before recirculation	

T2-18	M/H	The conditions of polymerization for these materials are concentration dependent and pH dependent. Changed ranking based on following email from Litman: "Due to my initial misinterpretation of the condition I want to change my ratings on both from Low to Medium. And if I can take a page from Apps's workbook both should be M-H"	PK	The concentrations are relatively low and the close to neutral pH do not provide optimal conditions for polymerization.
T2-19	L	During this phase coprecipitation is not likely to significantly contribute since the leaching of the non-metallics occurs over a longer period of time.	PK	The extent is not possible to predict accurately since coprecipitation not a thermodynamic property.
T2-20	H	Peroxide production will be immediate following the break since all hydrogen control will be lost. Nickel and cobalt will be easily solubilized, iron will precipitate as iron hydroxide.	K	PWR shutdown data base is an excellent source of information for this phenomenon.
T2-21	M/H	During this time frame, the leaching of the precipitating ions from non-metallic components has not yet become significant. Fibrous materials will begin to collect precipitated material but there will not be significant amount of deposition. Changed ranking due to following email from Litman: Due to my initial misinterpretation of the condition I want to change my ratings on both from Low to Medium. And if I can take a page from Apps's workbook both should be M-H.	K	Results of the ICET program.

T2-22	L	Galvanic corrosion effects during this early phase would not at all be significant.	K	Galvanic corrosion will not contr bute significantly ot the mass of precipitated material. However, with the significant mass of carbon steel, zinc and aluminum in containment, it is possible that electrical isolation of these components will be nullified eventually leading to long term component degradation/failure.
T2-23	H	The stagnant conditions that will exist outside the ZOI will promote settling of insoluble materials. Additionally, there are drop out areas and surfaces where precipitates can get trapped, minimizing their transport to the sump.	PK	Very dependent upon plant configuration and materials.
T2-24	L	The amount of organics introduced at this point is l kely to be low.	UK	Don't know immediate effects of specific organics on this combination of containment chemicals.

Table F-6: MacDonald T2 Ranking Summary

Issue Number	Importance (H, M, L)	Rationale	Knowledge (K, PK, UK)	Rationale

| T2-1 | M | Hydrogen, being a reducing agent, has a significant impact of the redox potential of the containment pool and of wetted surfaces. The redox potential, in turn, determines the valency of corrosion products and the rates of corrosion of many materials. Consequently, characterization of hydrogen sources within containment in the event of a LOCA is vital for predicting the chemical environment that will exist in the system over the 30s to t_{recirc} time frame. Initial rankings changed due to following email, rankings indicated separately: Please find attached a table with my re-assessments of some originally "high" ratings of issues. On examining these issues again, it was evident that I did not fully recognize the impact of:
1. Time of exposure after the initial LOCA.
2. The amount of the specific material in the containment (e.g., zinc).
My re-assessments also reflect the results of continuing radiolysis modeling, albeit with less-than-optimal codes that have been jerry-rigged from our reactor modeling work. Nevertheless, if the fields are as high as Bob indicates, the eventual production of nitric acid, in addition to the elevation of the redox potential, becomes a critical issue. | K | The principal sources of hydrogen are the inventory in the primary coolant prior to the break and corrosion of the highly reactive metals (e.g., Al and possibly Zn, but not Fe, Cu, Pb, or the stainless steels or nickel alloys). Hydrogen may also be generated by the Schikorr reaction from precipitated ferrous hydroxide [$3Fe(OH)_2 \rightarrow Fe_3O_4 + 2H_2O + H_2$] and from radiolysis, although the relative importance of these sources is difficult to predict in a general sense. Nevertheless, the kinetics of hydrogen release from these various sources are well-defined. |

T2-2	M	The injection of boric acid of 2800ppm concentration (B) is not likely to cause ant significant swings in pH of the containment pool and hence is not likely to significantly affect the corrosivity of the pool environment. A possible exception is if the boric acid solution evaporates in contact with hot fuel to form a concentrated, highly corrosive wet boric acid melt.	K	The impact that boric acid will have on the pH of the containment pool is readily calculated if the pool components prior to the addition can be reasonably defined.
T2-3	M	Containment spray will contact many surfaces after a LOCA and the impact that it will have on the corrosion of those surfaces may affect the total inventory of debris. However, the lowest pH (4.5-5) is similar to that of acid rain, and recognizing that many of the surfaces are protected by paint or other coatings, it is posited that little damage will occur in the time-frame of 30s to t_{recirc} time frame.	K	Extensive work has been reported on the corrosion of materials, including painted metals and alloys, under acid rain conditions, so that the inventory of corrosion products that might be generated from this source can be estimated. High pH spray (NaOH, pH =8.5-10.5) will not significantly impact Fe, Ni-based alloys, Cu, Pb, or alloys, such as the stainless steels. However, the upper end of the pH range may significantly enhance the corrosion of unprotected Al. Sufficient data are available in the literature to ascertain the magnitude of the impact.
T2-4	M	See T2-3.	K	See T2-3.

T2-5	M	Sodium tetra borate ($Na_2B_4O_7$) is a buffering agent that is released with ice in containment during a LOCA, with the specific purpose of pH-buffering the pool contents. However, much of the required buffering capacity is already provided by the coolant boric acid and the ECCS boric acid.	K	The impact that sodium tetra borate (STB) will have on the pH of the containment pool is readily calculated if the pool components prior to the addition can be reasonably defined.
T2-6	M	Tri sodium phosphate (Na_3PO_4) is a buffering agent that is released from baskets on the containment floor during a LOCA with the specific purpose of pH-buffering the pool contents. However, much of the required buffering capacity is already provided by the coolant boric acid and the ECCS boric acid.	K	The impact that tri sodium phosphate (TSB) will have on the pH of the containment pool is readily calculated if the pool components prior to the addition can be reasonably defined.
T2-7	H	The amount of debris that is washed into the containment pool will critically affect the tendency of the pump screens to clog.	UK	Depends on the design of the system, the inventory of "loose material to begin with, and the inventory of material produced during the first 30s of the LOCA. Likely to be difficult to predict with reasonable accuracy.
T2-8	M	Absorption of CO_2 from the containment atmosphere might lead to the precipitation of carbonate solids (e.g., $CaCO_3$). The inventory of $CaCO_3$ that might result is predicted to be several hundred kilograms, which could represent a sufficiently high amount to clog the filters.	K	The maximum possible inventory of $CaCO_3$ is readily calculated from the volume of air in containment and the concentration of Ca^{2+} in and the volume of the containment pool contents.

T2-9	H	Debris dissolution will determine the total inventory of solid material in the containment pool. Furthermore, the dissolved species may react and co-precipitate among themselves to form new, solid phases that were not originally in the pool.	PK	Because the pool environment is chemically very complex, and because dissolution experiments have generally been reported only for much simpler systems, it is not clear how applicable the current experimental database is to this problem.
T2-10	M	The creation of $CaCO_3$ nuclei for the precipitation of other solid phases is a matter of great importance in this early time frame, as precipitation is expected to contribute substantially to the total solid phase inventory in the pool. Initial rankings changed due to following email, rankings indicated separately: Please find attached a table with my re-assessments of some originally "high" ratings of issues. On examining these issues again, it was evident that I did not fully recognize the impact of: 1. Time of exposure after the initial LOCA. 2. The amount of the specific material in the containment (e.g., zinc). My re-assessments also reflect the results of continuing radiolysis modeling, albeit with less-than-optimal codes that have been jerry-rigged from our reactor modeling work. Nevertheless, if the fields are as high as Bob indicates, the eventual production of nitric acid, in addition to the elevation of the redox potential, becomes a critical issue.	PK	The precipitation of $CaCO_3$ has been studied extensively in the petroleum and marine industries, so that the basic physical chemistry of the formation of carbonate nuclei is well understood. However, the kinetics of formation of the nuclei are sensitive to the presence of various surfactants (e.g., organic acids and amines), which may derive from the decomposition of insulation, paint, and coatings, for example.

| T2-11 | H | Intense mixing enhances mass transfer effects, which, in turn, affects the rates of processes such as dissolution (including CO_2 solution), precipitation, and coagulation, all of which impact the total solids inventory in the containment pool. | PK | The basic impact of agitation and mass transfer on the kinetics of dissolution, solution, precipitation, and to a lesser extent coagulation, is fairly well understood in general terms, but their magnitudes are sensitive to geometrical and surface (e.g., presence of precipitates) effects. Accordingly, mixing effects tend to be highly specific to the system under consideration. The good news is that powerful finite element codes are available for predicting mass transfer effects in complex geometries. |

| T2-12 | L | Corrosion of concrete by boric acid will add to the dissolved species inventory in the pool and may lead to the spalling of solid concrete. The dissolved species may react with other species in the pool to precipitate additional solid phases that may clog the pump screens and the spalled concrete will add to the total mobile solids inventory in the pool. Initial rankings changed due to following email, rankings indicated separately: Please find attached a table with my re-assessments of some originally "high" ratings of issues. On examining these issues again, it was evident that I did not fully recognize the impact of: 1. Time of exposure after the initial LOCA. 2. The amount of the specific material in the containment (e.g., zinc). My re-assessments also reflect the results of continuing radiolysis modeling, albeit with less-than-optimal codes that have been jerry-rigged from our reactor modeling work. Nevertheless, if the fields are as high as Bob indicates, the eventual production of nitric acid, in addition to the elevation of the redox potential, becomes a critical issue. | UK | Little appears to be known of the corrosion of concrete by boric acid, in terms of either the kinetics or the solid phases that eventually form. |

T2-13	H	Radioactive iosotopes of Fe, Ni, and other metals that are ejected from the LOCA break in the form of spalled oxide particles will eventually find their way into the containment pool. The resulting radiolysis of water will lead to an increase the redox potential, which in turn will considerably affect the chemistry of the pool environment.	PK	While the radiolysis of water can be effectively modeled, provided that the dose rate can be defined, the processes whereby particles of oxide are transported within the pool are less well understood or defined. Previous work on the transport of silt in water supply systems may provide a useful basis for describing these processes.
T2-14	M	If NiO (or other oxides) catalyze the production of hydrogen from the radiolysis of water, it could have a significant impact on the redox potential. However, the stoichiometry of the overall reaction is such that catalysis should also occur in the production of oxidizing species, with the net effect possibly being moot.	UK	If NiO does catalyze the production of H_2 via the radiolysis of water, the effect on the redox potential can be readily calculated using the Mixed Potential Model. However, as search of the literature failed to yield any data or reported observations in support to this postulate.
T2-15	M	Complexing between metal ions and certain organic molecules (amines, acids, and heterocycles) is well-known and would generally increase the solubility of solids. Thus, solubility products may not be exceeded and hence certain solid phases may not form when they would form in the absence of complexing.	K	Large databases exist on the stability constants for a wide range of complexes between metal cations and organic species.
T2-16	M	Organic sequestration is the process whereby metal ions absorb into organic phases, most notably into ion exchange materials.	K	Large databases exist on the ion exchange capacities of a wide range of organic polymers containing ion exchange groups.

T2-17	M	Auxiliary component cooling line failure resulting in the release of oxidizing species (chromate, molybdate) and organic species (nitrites, hydrazine, tolytriazole, benzotriazole) could have a very significant effect of the chemistry of the containment pool and hence could lead to the formation of new solids. Initial rankings changed due to following email, rankings indicated separately: Please find attached a table with my re-assessments of some originally "high" ratings of issues. On examining these issues again, it was evident that I did not fully recognize the impact of: 1. Time of exposure after the initial LOCA. 2. The amount of the specific material in the containment (e.g., zinc). My re-assessments also reflect the results of continuing radiolysis modeling, albeit with less-than-optimal codes that have been jerry-rigged from our reactor modeling work. Nevertheless, if the fields are as high as Bob indicates, the eventual production of nitric acid, in addition to the elevation of the redox potential, becomes a critical issue.	PK	In principle, the impact that oxidizing and reducing agents will have upon the redox potential and hence upon the tendency for high valency precipitates to form [e.g., $Fe(OH)_3$ versus $Fe(OH)_3$] can be estimated using the mixed potential model (MPM), but only if exchange current density and transfer coefficient data are available for the redox reactions. Corrosion rates may also be effected, but we note that the triazoles are effective corrosion inhibitors, particularly of the corrosion of copper and copper alloys.

T2-18	H	Polymerization is the process by which hydrolyzed cations form a precipitated solid, generally via colloids. This process of precipitation is well-documented and understood and is expected to be a principal mechanism for the formation of solids in the containment pool.	PK	Extensive data are available for the hydrolysis and precipitation of solids from solutions containing simple cations, such as Al^{3+} and Fe^{3+}, but few data are available for more complex systems (e.g., precipitation of alumino-silicates). Provided that chemical potential data for the initial state ("reactants") and the final state (the precipitate) are available, the Gibbs energy minimization technique for calculating the system composition at equilbrium is most useful, because the properties of intermediate species need not be known.

| T2-19 | M | Co-precipitation is the process whereby a carrier precipitated phase carries down species that would not ordinarily form precipitates under the prevailing conditions. Co-precipitation can greatly increase the volume of a precipitate that may form and hence could lead to clogging of the pump screens. Initial rankings changed due to following email, rankings indicated separately: Please find attached a table with my re-assessments of some originally "high" ratings of issues. On examining these issues again, it was evident that I did not fully recognize the impact of:
1. Time of exposure after the initial LOCA.
2. The amount of the specific material in the containment (e.g., zinc).
My re-assessments also reflect the results of continuing radiolysis modeling, albeit with less-than-optimal codes that have been jerry-rigged from our reactor modeling work.
Nevertheless, if the fields are as high as Bob indicates, the eventual production of nitric acid, in addition to the elevation of the redox potential, becomes a critical issue. | PK | Although co-precipitation has been extensively studied in analytical chemistry, radio chemistry, and in product purification and separation technology, the work is largely empirical in nature and has been performed on relatively simple systems (e.g., co-precipitation of radium with barium sulfate). The theoretical base is insufficiently well-developed to predict co-precipitation phenomena in the complex systems of interest in a LOCA. |

T2-20	H	Water radiolysis produces a variety of oxidizing (O_2, H_2O_2, OH) and reducing (H_2, H, O_2^-, etc) species in non-equilibrium concentrations. These species establish the redox potential of the environment, which determines the tendency for oxidation and reduction reactions to occur. Accordingly, water radiolysis can have an important impact of the chemistry of the pool.	K	The radiolysis of water has been modeled extensively in nuclear reactor technology to estimate the concentrations of the principal radiolysis products. These concentrations can then be used in Mixed Potential Models to estimate the redox potential and the corrosion potentials of various metals and alloys. The corrosion potentials can then be used to estimate the general corrosion rates and the suscept bilities of the metals and alloys to localized corrosion processes using the Point Defect Model.
T2-21	H	Inorganic agglomeration is the process whereby colloidal particles (1-100nm in diameter) that are formed by cation polymerization (see T2-18) come together to form larger particles and eventually precipitates.	PK	Destabilization of colloidal systems ("solutions"), usually by increasing the ionic strength of the solution, which shrinks the thickness of the diffuse layer and allows attractive van der Waal forces to cause the particles to agglomerate, is a well-understood phenomenon. However, it is also physically complex, particularly for the multi-component systems of interest in this study. Even with an extensively developed theoretical base, it is doubtful that meaningful predictions can be made.
T2-22	M	Galvanic coupling between metals that are widely separated in the electrochemical series (e.g., Al-Cu, Al-stainless steel) enhances the corrosion rate of the more active component.	K	Galvanic corrosion has been studied extensively and is well-understood. The process can be modeled using a variant of the Mixed Potential Model and the corrosion rates of both components of the couple can be effectively calculated.

T2-23	H	Deposition and settling are the final processes in the precipitation process. Settling occurs when the mass of a particle becomes so large that it cannot be maintained in suspension by Brownian motion.	K	By knowing the particle density and volume, and hence mass, it is possible tp predict the maximum size that will remain in suspension. However, various surfactants, some of which may be present in the pool, are known to prevent settling/deposition. The scientific literature on soaps and detergents provides a wealth of information on this subject.
T2-24	H	Organic agglomeration is the process whereby organic colloidal particles (1-100nm in diameter) that are formed by hydrophilic micellation (cf T2-21) come together to form larger particles and eventually precipitates.	PK	Destabilization of organic colloidal systems ("solutions"), usually by decreasing the pH of the solution, so as to protonate the acid groups and hence annihilate the repulsive surface charge, allows attractive van der Waal forces to cause the particles to agglomerate. This process is a well-understood phenomenon. However, it is also physically complex, particularly for the multi-component systems of interest in this study. Even with an extensively developed theoretical base, as with inorganic agglomeration (T2-21), it is doubtful that meaningful predictions can be made.

APPENDIX G

PIRT Panelist Evaluations of T3 Phenomena:

ECCS Recirculation from onset of Recirculation until

24 hours into LOCA

Table G-1: ECCS Recirculation (t_{recirc} to 24 hours)[a,b]

Issue Number	Phenomena	Description	Implications
T3-1	TSP pH control	Contained in basket on containment floor that dissolve	Slightly alkaline containment pool environment (pH = 7 – 8); all TSP dissolved within 1 to 4 hours post LOCA.
T3-2	NaOH pH control	NaOH injected through containment sprays.	Alkaline containment pool environment (pH = 8.5 – 10.5); pH buffering completed typically within 4 hours.
T3-3	STB pH control	Released with ice in containment during a LOCA	Alkaline containment pool environment (pH = 8 – 9); all STB should be introduced within a few hours.
T3-4	NaOH Injection	NaOH, B laden spray impinges on material surfaces	1. Solution may accelerate corrosion of metallic/non-metallic surfaces due to high pH, boron concentration. 2. Solution may inhibit corrosion by formation of an initial passive layer impervious to future corrosion.
T3-5	Cable degradation	The insulation surrounding electrical cables located throughout containment is affected by radiolysis, releasing chlorides.	The chlorides affect radiolysis and ionic strength of containment pool environment.
T3-6	Radiolytic environment	Radiolysis reaction changes solution's redox potential; H_2, O_2, and H_2O_2 balance; peroxide formation occurring during this time frame.	Could affect chemistry in containment pool which affects species which form - most important for determining redox potential (e.g., Hanford tank: mixture of different components @ different phases)
T3-7	F berglass leaching	Fiberglass-based insulation products begin leaching constituents; e.g., Si, Al, Mg, Ca, etc.	Dissolution source term for producing ions which contr bute to containment pool chemistry.
T3-8	Secondary system Contamination	SG tube rupture leads to secondary side chemicals being added to containment pool.	1. Possible additional source for organic materials which could delay precipitation through complexation or by sequestering metal ions. 2. Could accelerate material corrosion.
T3-9	Flow-induced nucleation[c]	Nucleation sites created from cavitation, deareation, air entrainment, turbulence (attachment opportunity)	Increases possibility of precipitation/chemical species formation

T3-10	Turbulent mixing[c]	Turbulence in containment pool (such as directly under the break) either inhibits or promotes precipitate formation and growth	1. Could inhibit growth due to turbulent shear forces which don't allow solid species to agglomerate. 2. Could promote formation and growth by providing more opportunity for particles to collide and overcome diffusion. 3. Affects the solid species concentration within the containment pool which could contribute to head loss & other ECCS performance degradation.
T3-11	Quiescent settling of precipitate[c]	Quiescent flow regions within containment pool promote settling	Little flow allows larger size, more stable particles/precipitates to form which promotes settling.
T3-12	Electrostatic scavenging	Material surfaces electrostatically attract particles and lead to deposition/agglomeration on the surface; thin accumulation limited to 2 to 3 layers of material. It is a function of solution PZC and surface and particle charges.	1. Provides a mechanism for retaining chemical products either in benign areas (e.g., walls, surfaces) or in areas that may impact system performance (e.g., sump screen, pipe internals). 2. Surface layer either inhibits or enhances material corrosion.
T3-13	Chemically induced settling	Chemical species attach to or coat particulate debris which leads to settling. Examples are Al coating on Nukon fiber shifting the PZC, or formation of a hydrophobic organic coating	Results in less particulate debris and chemical product transporting to and either accumulating on or passing through the sump screen.
T3-14	Agglomeration & Coagulation	Formation of larger clumps of smaller particulates: Depends on PZC; ionic strength (the higher the strength the smaller the distance for agglomeration); sensitive to many factors including shape factors, maximum particle size.	1. May be necessary to form large enough particles to result in tangible effects on ECCS performance 2. Can occur quickly if conditions are right. 3. Existence of organic species can decrease likelihood
T3-15	Particulate nucleation sites	Particles within containment create nucleation sites for chemical precipitation: examples include radiation tracks; dirt particles; coating debris; insulation debris; biological debris, etc.	Environment is created which foster formation of solid species that could lead to ECCS degradation.

T3-16	Additional debris bed chemical reactions	Concentration of radionuclides (100's of Curies available) acts as a "resin bed" or chemical reactor. A number of possible reactions occur.	1. Hydrogen peroxide formation 2. $Fe^{+2} \rightarrow Fe^{+3}$ increase due to increase in redox potential; 3. Organic materials decompose 4. possible coating of NUKON to reduce solubility; glass embrittlement (shorter fiber);
T3-17	Sump screen: high localized chemical concentrations	High localized concentrations of certain species drive the reaction efficiency at the sump screen	1. Effect could be at the exterior surface of insoluble materials, the internal interstitial void material, or both 2. Possible effect on the chemical species which form within the sump screen bed.
T3-18	Sump screen: fiberglass morphology	Localized chemistry alters the fiberglass contribution to the chemical environment	1. Possible coating of fiberglass fibers reduces leaching/solubility. 2. Radiation leads to glass embrittlement; shorter fibers form which can pack more densely (increase head loss) and provide additional surface area for leaching/reactions.
T3-19	ECCS pump: seal abrasion	Abrasive wearing of pump seals (e.g., magnetite - high volume/concentration of mild abrasive) creates additional materials that contribute to sump pool chemistry.	1. Additional particles that may contribute to reactor core clogging 2. Particles may add additional sump screen loading. 3. Particles may affect chemical species formation.
T3-20	Heat exchanger: solid species formation	Concentrations/species that are soluble at the containment pool temperature precipitate at lower ($\Delta T \sim 15\text{-}20C$) heat exchanger outlet temperature.	1. Species remain insoluble at higher reactor temperatures and affect ability to cool the reactor core. 2. Species remain insoluble at higher containment pool temperatures and cause additional head loss upon recirculation.
T3-21	Heat exchanger: deposition & clogging	Solid species which form in the heat exchanger lead to surface deposition and/or clogging within closs-packed head exchanger tubes (5/8" diam.)	1. Severe clogging/deposition causes flow decrease through heat exchanger core and an inability to cool reactor core. 2. Less severe deposition may degrade heat transfer and degrade heat flow from the reactor core.

ID	Phenomenon	Description	Effects
T3-22	Reactor core: fuel deposition	ΔT increase (+70C from sump pool) and retrograde solubility of some species (e.g., Ca silicate, Ca carbonate, zeolite, sodium calcium aluminate) causes scale, build-up on reactor core	1. Decreases heat conduction values from fuel. 2. Localized boiling occurs due to insufficient heat removal. 3. Deposits spall off, creating additional debris source.
T3-23	Reactor core: hydrogen increases	H_2 builds up in from cladding oxidation	1. Affect redox potential and chemical environment. 2. Increases containment hydrogen concentration; may lead to conflagration.
T3-24	Reactor core: diminished heat transfer	Insulation debris and chemical products mixed within water cause a reduction in the effective heat transfer capabilities (C_p) of mixture. The precipitation of this material is initiated in RHR heat exchanger.	Ability to remove heat from the fuel is diminished.
T3-25	Reactor Core: Blocking of flow passages	Chemical products, in combination with other debris, blocks primary flow passages for getting cooling water through core	Heat transfer may be impeded; flow is forced to bypass the lower plenum debris screens
T3-26	Reactor Core: Particulate settling	Particulate settling occurs due to relatively low, upwards flow (for cold leg injection) within reactor	Compacted deposits forms which may impede heat transfer and water flow, especially for lower portions of reactor fuel.
T3-27	Reactor Core: Precipitation	ΔT increase (+70C from sump pool) and retrograde solubility of some species (e.g., Ca silicate, Ca carbonate, zeolite, sodium calcium aluminate) causes precipitation, additional chemical product formation	1. Products contribute to debris in reactor core which blocks flow passages, impedes heat transfer. 2. Additional precipitate is created which travels to the sump screen and leads to head loss.
T3-28	Exposed, Uncoated Concrete Dissolution	Dissolution can result in pH increases (depending on surface area) up to 10.5 below concrete surface relatively quickly.	1. This could result in pH increase in containment pool over time. 2. Leaching of concrete elements (e.g., Ca, Si, etc.) into containment pool.
T3-29	Coatings Dissolution	Dissolution/leaching of epoxy, a kyd, or zinc-based coatings and primers occurs	1. Additional source term for products affecting the containment pool chemistry. 2. Chlorides affect radiolysis and ionic strength of containment pool environment.

T3-30	Boric Acid Corrosion	Concentrated boric acid pooled on unsubmerged portions of large ferritic steel components (e.g., piping RPV, pressurizer, steam generator) continues to cause corrosion of these materials.	1. Contribute additional Fe ions which may be transported to containment pool. 2. Could affect coprecipitation and solid species concentrations 3. May lead to long-term structural weakness if severe.
T3-31	CO_2/Carbonate Radiolysis	Radiolysis of CO_2 and carbonates could lead to the formation of organic acid ligands (e.g., acetate, butyrate, oxalate) This is observed during the leaching of spent fuel by groundwaters containing bicarbonate ions	1. Phenomena would complex metal species in solution, thereby enhancing dissolution rates and diminishing the precipitation of phases containing these complexed species. 2. Could occur within containment pool, reactor vessel, or within the sump screen debris bed.

[a] Sump pool temperatures are expected to range between 50 – 90C during this time period.
[b] Phenomena listed in earlier time periods (i.e., T1, T2) may continue to be active or important during this time.
[c] Bulk flow velocities = 0.005 ft/s to 0.1 ft/s (0.15 – 3 cm/s)

Table G-2: Apps T3 Ranking Summary

Issue Number	Importance (H, M, L)	Rationale	Knowledge (K, PK, UK)	Rationale
T3-1	L	Rate of dissolution of TSP will have only a transient affect on coolant chemistry.	PK	Effect would be specific to the type of LOCA and plant design.
T3-2	H	Major effect on coolant composition.	K	Plant operating parameters well known.
T3-3	H	Major effect on coolant composition.	K	ICET series provides some information.
T3-4	L/M	Effect is transient.	PK	ICET series provides some information.
T3-5	L	Quantities are small and the system is pH buffered.	PK	Extent of degradation can be estimated.
T3-6	L/M	Some impact on corrosion.	PK	Specific impact on corrosion requires study.
T3-7	M	Fiberglass leaching is not extensive during the time frame in question.	K	ICET series and supplementary lab studies provide quantitative information.
T3-8	L/M	Volume of contaminants not expected to be large.	UK	Impact would be LOCA and plant specific. Effect of chemistry has not been evaluated.

T3-9	L/M	Effect depends on the extent of reaction and supersaturation. Likely to accelerate rates rather than change eventual outcome.	UK	Impact would be LOCA and plant specific. Additional study would be required.
T3-10	L/M	Effect depends on the extent of reaction and supersaturation. Likely to accelerate rates rather than change eventual outcome.	UK	Impact would be LOCA and plant specific. Additional study would be required.
T3-11	M	Precipitates would not affect sump screen performance.	UK	Impact would be LOCA and plant specific. Additional study would be required.
T3-12	L	Total amount of material immobilized would be small. Could favorably affect passivation.	UK	Impact would require evaluation and experimental study.
T3-13	M/H	Adsorption or coating could inhibit further reaction.	PK	Experiments have already been performed in support of the ICET series
T3-14	M/H	Could impact sump screen blockage.	PK	Experiments have already been performed in support of the ICET and head loss test series
T3-15	M	Effect is important, but will happen in any case.	UK	Very difficult to quantify. Uncontrolled.
T3-16	L/M	Overall impact not expected to be great.	UK	Magnitude and extent of the effect is unknown, and would require further study.
T3-17	L	Chemical reactions within the sump screen bed are expected to be secondary.	UK	Would require further study.

T3-18	L	Effect is likely to be minor during the initial 24 hr.	PK	ICET series provides some indications of the significance of this phenomenon.
T3-19	L	Amount of material generated would be very small. Pump design should prevent this from happening, or at least mitigate any effect.	UK	Would require characterization.
T3-20	M	Potentially adverse effect on cooling.	PK	Effect could be calculated, but experimental verification would be required.
T3-21	M/H	Potentially serious adverse effect on cooling.	PK	Effect could be calculated, but experimental verification would be required.
T3-22	M/H	Potentially serious adverse effect on cooling.	PK	Effect could be calculated, but experimental verification would be required.
T3-23	L	Cooling decreases the rate of hydrolysis. Hydrogen would partition into the containment-building atmosphere become relatively inert.	PK	Kinetics should be available to calculate the effect as a function of temperature.
T3-24	M/H	Adverse effect on core cooling, and potential damage.	UK	Effects depend on many unknown factors. Tests and modeling would be required.
T3-25	M/H	Adverse effect on core cooling, and potential damage.	UK	Effects depend on many unknown factors. Tests and modeling would be required.
T3-26	M/H	Adverse effect on core cooling, and potential damage.	UK	Hydrodynamic calculations would permit estimates to be made of material settling in terms of amount and location.

T3-27	M/H	Adverse effect on core cooling, and potential damage.	PK	Kinetics of precipitation should be available to allow calculation of amount.
T3-28	L	Effect minimized by diffusion and low exposed surface area. pH is buffered.	PK	Information to quantify this effect is probably available in the literature.
T3-29	L	Most coatings will be resistant to dissolution. Dissolved products would be minor in amount.	PK	Information to quantify this effect is probably available in the literature.
T3-30	L/M	Products unlikely to have a major effect on corrosion or sump screen blocking. Unlikely to cause structural damage within first 24 hr.	PK	ICET series provide some information.
T3-31	L	Effect would be small because of relatively low carbonate concentration.	PK	Information on radiolysis of carbonate probably available to allow calculation of the magnitude of the effect.

Table G-3: Chen T3 Ranking Summary

Issue Number	Importance (H, M, L)	Rationale	Knowledge (K, PK, UK)	Rationale
T3-1	M		K	
T3-2	M		K	
T3-3	M		K	
T3-4	L		PK	
T3-5	No comment		No comment	
T3-6	M		PK	
T3-7	M		K	
T3-8	L		PK	

T3-9	L		PK	
T3-10	M		PK	
T3-11	H		K	
T3-12	L		PK	
T3-13	L		PK	
T3-14	H		PK	
T3-15	L		UK	
T3-16	L		UK	
T3-17	L	too low a residence time	PK	
T3-18	L	should be the same anywhere in the sump	PK	

T3-19	L	not likely to add significant amount of solids into the system	K	
T3-20	L	not enough time	K	
T3-21	L	not enough time	K	
T3-22	L	not enough time	K	
T3-23	No comment		No comment	
T3-24	L	does not affect solid formation	K	
T3-25	M	if a piece of large debri goes into core	K	
T3-26	L		K	
T3-27	M		PK	
T3-28	M		PK	

T3-29	L	unlikely to form solids	UK	
T3-30	L		PK	
T3-31	No comment		No comment	

Table G-4: Delegard T3 Ranking Summary

Issue Number	Importance (H, M, L)	Rationale	Knowledge (K, PK, UK)	Rationale
T3-1	H	The trisodium phosphate is a large contributor to solutes. Additional rationale provided in email: The effects on solids formation are demonstrably large, and within this ~30 second to 30 minute time frame, as the phosphate combines with calcium to form the calcium phosphate precipitate that the "Head Loss Testing" found to be so effective in stopping flow.	K	Studied by ICET and other testing.
T3-2	H	The NaOH is large contr butor to pH balance and solute loading.	K	Studied by ICET and other testing.
T3-3	H	The sodium tetraborate is a large contributor to solutes.	K	Studied by ICET and other testing.
T3-4	M	Shown by ICET and other testing to contribute to aluminum corrosion. I downgraded Delegard from high to medium based on email stating that this issue is secondary in light of the others and/or because it has been considered in other time frames.	K	Studied by ICET and other testing.
T3-5	M	Cable (insulation) important under thermal and radiolytic conditions.	UK	Needs testing; see T2-15 and T2-24.
T3-6	H	Of most importance for organic materials.	UK	Needs testing. See T2-20.

T3-7	M	ICET and other testing showed these effects.	K	Studied by ICET and other testing.
T3-8	M	See T2-17. The components in the cooling lines may have a significant effect but their concentrations are small.	PK	Testing needed to see if the components have an effect. The effects on chromate and molybdate on aluminum corrosion could be large.
T3-9	L	The conditions of flow to affect nucleation were tested by ICET.	K	Studied by ICET.
T3-10	M	The effects of turbulence were tested by ICET. See T2-11.	K	Studied by ICET.
T3-11	M	This does not lend itself to chemical effects testing.	UK	This phenomenon could be exploited by design of, effectively, cascade settlers in the floor of the containment building. This, however, seems to be outside the scope of the chemical testing.
T3-12	L	These phenomena (electrostatic scavenging, chemically induced settling, agglomeration/coagulation, and particulate nucleation sites) are potential explanations of observed behaviors but are not, in themselves, unexamined chemical / physical variables in the post-LOCA environment.	PK	These phenomena and mechanisms should be considered to explain the observations from testing.
T3-13	L		PK	
T3-14	L		PK	
T3-15	L		K	

T3-16	M	Radiation fields at the sump screen could be enough to harden and crosslink organic materials.	UK	Testing of the effects of high radiation fields needed for sump screen beds. Effects expected to be higher for the organic materials.
T3-17	M	These effects were studied in ICET and head loss tests.	PK	Such conditions were investigated in ICET and head loss testing.
T3-18	M	This is a subset of T3-16.	UK	Testing in radiation field needed.
T3-19	H	The contribution of solids eroded from the pump seals to the total solids budget is minor. However, the failure of the recirc pumps would be disastrous.	UK	The effects of the recirculating suspended solids on the integrity of the pumps require testing or at least evaluation by pump engineers.
T3-20	H	Effects of solids deposition (by solubility and caking) on cool surfaces potentially can build large particles for transport and blockage of sump screen.	UK	The effects of cooler surfaces on the behavior of the recirculating coolant were not examined in testing to-date.
T3-21	H		UK	
T3-22	H		UK	
T3-23	L	H_2 disengages rapidly from hot water and is unlikely to affect redox chemistry greatly. See T2-1.	PK	Real issue is overall Eh of system and this has not been measured under radiolytic conditions.
T3-24	H	These three phenomena could affect fuel integrity and cause fuel cladding failure. This would increase solids loading in the coolant system and hence blockage of the sump screen as well as increase the radiolytic exposure to the post-LOCA environment. The UO_2, should it escape from the fuel cladding, is also very hard and abrasive and could cause earlier failure of pump seals and bearings.	UK	Engineering evaluations may be sufficient to assess the impact of solids deposition and loading of the core thermal and hydraulic transfer.

T3-25	H		UK	
T3-26	H		UK	
T3-27	H	Effects of solids deposition (by retrograde solubility, solvent evaporation, or caking) on hot core surfaces potentially can build large particles for transport and blockage of sump screen. Note that borate deposition also can increase metal attack.	UK	The effects of hotter surfaces on the behavior of the recirculating coolant were not examined in testing to-date.
T3-28	M	This phenomenon, though important, has been considered in ICET and other testing.	K	Studied by ICET and other testing.
T3-29	M	The decomposition of coatings (e.g., paints and electrical insulation) by thermolysis and radiolysis, can steeply exacerbate solids loading on the sump screens. See T2-24. I downgraded Delegard from high to medium based on email stating that this issue is secondary in light of the others and/or because it has been considered in other time frames.	UK	Testing is needed here for both raw organic materials (such as lubricating oils) and thermally and radiolytically degraded organic materials (e.g., paints, plastics) to assess their potential to agglomerate small sized solids.
T3-30	M	Borates at high temperatures are used to digest minerals in chemical assay. Their effects on hot and unwetted metal could be severe.	PK	The attack of borates on unwetted metal coupons should be tested.

| T3-31 | L | Free carbonate concentrations likely will be low in this calcium-rich coolant system. | UK | Testing of this mechanism can be part of the larger suite of tests on the effects of radiolysis in the coolant system. |

Table G-5: Litman T3 Ranking Summary

Issue Number	Importance (H, M, L)	Rationale	Knowledge (K, PK, UK)	Rationale
T3-1	L	TSP will provide adequate buffering capacity for the entire fluid volume within about 2 hours.	K	Based on the system design. The recirculation flow rates and volumes provide for a 400,000 gallon turnover at a minimum of 6,000 gpm of two volumes in two hours.
T3-2	L	pH in the spray will be at pH 9.5 immediately upon commencement of spray. Water in sump will be at pH of 9.5 within the next hour (even before recirculation begins).	K	Based on the system design.
T3-3	L	pH in spray may be delayed a few minutes until sufficient water has passed through the ice condenser to dissolve all the STB. After that pH =8.5	K	Based on the system design.
T3-4	M	Solution will be at pH 9.5 will be warm (120-160F) and will be replenished rapidly based on the washing effect of the spray	K	Based on the system design
T3-5	L	The cable trays will start to be exposed to high levels of radiation at this point. However, degradation to chlorides is not instantaneous and will occur slowly over the next several days.	K	Cable insulation does undergo radiolysis to yield chlorides; eventually.
T3-6	M	Solubility of metal ions will be affected by the presence of peroxide.	K	PWR shutdown chemistry programs.
T3-7	M	The leaching process is time controlled. The actual leached concentrations will not yet be significant.	K	ICET program provides evidence that maximum leached concentration do not occur until days later.

T3-8	L	A steam generator tube rupture would not start to have an effect on the RCS and containment until RCS 'backfill' begins which is late in this time period.	K	SG make up is with CST water; this is basically demineralized water. All PWRS use All volatile treatment(AVT) thus during the first few minutes of the event the chemicals in the SG would be boiled off with secondary steam and into the condenser or to the turbine building roof.
T3-9	M	Linear flow rates are low. Any surface, which can 'trap' insoluble materials, will be able to do so. There is a lot of surface area over which the recirculated water will flow in containment before it reaches the sump.	UK	This will be a plant specific and event specific parameter.
T3-10	L	The affect of turbulence beneath the break may be significant. However the transit time to the sump would allow for re-agglomeration of any suspended material.	UK	This will depend on the break size, the flow out the break, and the precipitate formed.
T3-11	M	The containment floor in PWRs is a maze of components bolted to the concrete. The pathway from the break to the containment sump is not uniform, unidirectional flow. This type of path provides more significant opportunity for solid phases to agglomerate and settle. The flow rate in linear ft/second is also very small. This also provides additional time for settling.	PK	The quiescent portions of the flow path to the sump will be dependent upon the individual plant design, the location of the major precipitate contributors with respect to the sump, and the type of insulation that was affected during the event.
T3-12	L	All containment surfaces will be wetted. The only significant electrostatic charge would most likely result from galvanic corrosion effects. Those would be minimal during this time frame.	UK	This is a very general possibility. The PZC of any in place precipitates or surfaces would have some effect on the attraction of precipitates but the mass they would attract due to this phenomenon is likely to be small.

T3-13	L	There are a limited number of other materials available that will be transported in the post-LOCA environment	PK	The ratio of particle surface area to the amount of material that cold potentially coat particles is largely unknown. Some materials could be organics or oils.
T3-14	M	Precipitate aging during this time period will be in its early stages.	PK	This will be very dependent upon the type of precipitate, which in turn is dependent upon the type o insulation affected and the containment spray buffering agent.
T3-15	M	Dislodged CRUD from the reactor trip will provide a significant source of nucleation sites. These particles will range in size from <1 micron to >100 microns. Their density will be on the order of 4-6 g/cc.	PK	The specific effects of these materials on the insulation/buffering agent precipitate are as yet unknown.
T3-16	M	As the debris bed on the screens or filters increases, the level of radionuclide activity will increase. This will create an oxidizing environment (due to oxygen saturation) which will transform reduced species into oxidized species.	PK	Tests using radiation sources with precipitated materials need to be performed to better establish the effects of radiation fields on these materials.
T3-17	M	The debris bed will initially concentrate those materials which are particulate (fiberglass fibers, shards of insulation jackets or Cal-Sil, metal and coating fragments) and those which are insoluble precipitates resulting form chemical reactions.	PK	The known precipitates that will form are capable of trapping other materials as well as co precipitating other soluble ionic solids.

T3-18	M	F berglass consists of several different mineral silicates. In an intense radiation field, these silicates will be transformed into soluble silicate ions. These new silicate ions will be transported to various containment areas where new silicate precipitates may form.	K	These are know reactions of silica in radiation fields. The precipitation effects of containment spray buffering agents has already been established through the ICET program.
T3-19	H	Transport of significant quantities of CRUD trapped in agglomerated precipitates will occur. Some of these materials will get through screens and find their ways to the RHR pumps and safety injection isolation valves..	K	The effects of these types of particulates are known. Most pump seal faces are hardened surfaces to withstand normal service, which does not include high levels of abrasive, particulate matter.
T3-20	H	The recirculated water will have relatively high concentrations of ions due to leaching of insulation and other solid materials in containment. This initial solubilization will have occurred at temperatures much higher (from 50 to 200 F) than the RHR heat exchanger tubes surfaces.	K	Solubility of most of these materials is a function of temperature. The exceptions are some alumino-silicates and AlOOH. These two have retrograde solubility.
T3-21	L	The degree of heat exchange capacity loss will depend to great deal on the type of precipitated material. However, during the first 24 hours this should be negligible due to design margins inherent in the heat exchanger design.	K	Plant design characteristics and conservativisms put on heat exchanger capacities.
T3-22	L	During this time period the deposition of materials with retrograde solubility will begin. However it should not occur to a significant extent.	K	Kinetic controlled process

T3-23	L	The greatest danger to core clad oxidation is during the early phases of the event when core uncovery is more likely and fuel centerline temperature is at its peak.	K	Design and operating characteristics of PWRS.
T3-24	L	During this time period the deposition of materials with retrograde solubility will begin. However it should not occur to a significant extent. However if significant debris is transported through the sump screens that could block flow in fuel channels, this could lead to selected fuel assembly overheat conditions.	K	Fuel rod design characteristics.
T3-25	L	See T3-24	K	See T3-24
T3-26	L	During cold leg injection (up to 24 hours) the bottom of the reactor vessel will be location for the deposition of solid materials. Flow direction changes 180° at the vessel bottom, which will cause denser particulates to be deposited there.	K	Flow design of reactor core.
T3-27	L	Retrograde solubility materials will begin to deposit on fuel surfaces.	PK	No modeling has been done in any tests to determine what the extent of this effect is.
T3-28	L	The surface area of uncoated concrete will generally be small and the effect will be small during this limited contact time period.	K	Plant design. Leach rate characteristics of concrete. Leaching of basic compounds from concrete will be buffered by presence of boric acid.

T3-29	M	There are a variety of different coating materials. Their resistance to leaching and dis-bondment will depend on the their original application and qualification process.	PK	Different combinations of coatings exist at each plant. No concerted effort has been made yet to quantify what the dissolution characteristics of these coatings are.
T3-30	M	Boric acid corrosion of carbon steel is a significant industry issue. This has the potential to generate significant quantities of ferric hydroxide.	K	Industry experience with boric acid corrosion of carbon steel and mild steel components.
T3-31	L	Carbon dioxide/carbonates are limiting reagents in the precipitation reactions. The equilbrium process for precipitation of these ions will far outweigh the radiolytic effects of their decomposition to form organic acids. The terminal radiolysis reaction of organic acids reverts to carbonates.	PK	No modeling or testing of this particular radiolytic phenomenon with the precipitated materials has yet been accomplished.

Table 6: MacDonald T3 Ranking Summary

Issue Number	Importance (H, M, L)	Rationale	Knowledge (K, PK, UK)	Rationale
T3-1	M	See T2-6.	K	See T2-6.
T3-2	M	See T2-4	K	See T2-4.
T3-3	M	See T2-5.	K	See T2-5.
T3-4	M	Impingement of boron laden, NaOH spray on metallic and non-metallic components in containment may enhance corrosion. Assuming that the pH is in the range of 8.5-10.5, the metal that will be most likely affected is Al and then only at the upper extreme of this pH range.	PK	The corrosion of Al and other metals in mildly alkaline solutions have been extensively studied and data are available for estimating the corrosion rate. Likewise, information is available on enhancement of corrosion rate due to liquid jet impingement. The situation with regard to non-metallic materials is less well-defined, but most plastics and paints are stable when in contact with weakly alkaline solutions (but not when in contact with highly alkaline solutions).

T3-5	M	Radiolytic degradation of electrical insulation is known to release chloride ion, which may induce passivity breakdown on various metals and alloys (e.g., Al, stainless steel), thereby greatly enhancing corrosion rates and the formation of corrosion products. Initial rankings changed due to following email, rankings indicated separately: Please find attached a table with my re-assessments of some originally "high" ratings of issues. On examining these issues again, it was evident that I did not fully recognize the impact of: 1. Time of exposure after the initial LOCA. 2. The amount of the specific material in the containment (e.g., zinc). My re-assessments also reflect the results of continuing radiolysis modeling, albeit with less-than-optimal codes that have been jerry-rigged from our reactor modeling work. Nevertheless, if the fields are as high as Bob indicates, the eventual production of nitric acid, in addition to the elevation of the redox potential, becomes a critical issue.	K	The radiolytic degradation of plastic electrical insulation (e.g., Poly Vinyl Chloride, PVC) has been extensively studied and the rate of degradation can be estimated if the dose rate is known. The impact of chloride ion on passivity breakdown of metals and alloys can be calculated using the Point Defect Model.
T3-6	H	See T2-20.	K	See T2-20.

T3-7	M	F berglass thermal insulation leaching is expected to add to the inventory of dissolved species and to a decrease in the inventory of solids that might clog the pump screens. However, the dissolved products (Si, Al, Mg, Ca, etc. cations or oxyanions) may react with other components in the pool to form voluminous, gel-like products that would have the potential for blocking the screens.	K	The leaching behavior of fiberglass has been extensively studied and it appears not to be significant over a 24-hour period unless the pH > 10. However, little is known of the reactions of leachants with other components in the containment pool or of the impact that these species might have on the pool chemistry, in general.
T3-8	M	Rupture of steam generator tubes would allow the discharge of secondary-side coolant into containment. The secondary coolant is generally highly reducing in nature with hydrazine (NH_2NH_2) as the reducing agent and an organic amine (e.g., morpholine) or ammonia as the AVT (All Volatile Treatment) pH control protocol.	K	The secondary chemistry of a PWR is very well-defined with models being available for calculating important properties, such as pH, conductivity, etc. Gibbs energy minimization codes are also available that could be used to estimate the pH and composition of primary/secondary mixed systems that would result in the event of steam generator tube failures.

T3-9	L	Flow induced nucleation is the process whereby turbulence in a liquid or solution induces the formation of tiny bubbles through cavitation. These bubbles act as nucleation sites for precipitation of solids from super saturated solutions. Initial rankings changed due to following email, rankings indicated separately: Please find attached a table with my re-assessments of some originally "high" ratings of issues. On examining these issues again, it was evident that I did not fully recognize the impact of: 1. Time of exposure after the initial LOCA. 2. The amount of the specific material in the containment (e.g., zinc). My re-assessments also reflect the results of continuing radiolysis modeling, albeit with less-than-optimal codes that have been jerry-rigged from our reactor modeling work. Nevertheless, if the fields are as high as Bob indicates, the eventual production of nitric acid, in addition to the elevation of the redox potential, becomes a critical issue.	K	The basic physics is known and a variant of the method, sonification, is used as a means of inducing precipitation in industrial and laboratory systems.

T3-10	L	Turbulence in two phase systems is characterized by high shear forces being imparted to particles that tend to oppose agglomeration. In some systems, it may lead to an increased collision frequency between particles and hence may promote precipitation. Initial rankings changed due to following email, rankings indicated separately: Please find attached a table with my re-assessments of some originally "high" ratings of issues. On examining these issues again, it was evident that I did not fully recognize the impact of: 1. Time of exposure after the initial LOCA. 2. The amount of the specific material in the containment (e.g., zinc). My re-assessments also reflect the results of continuing radiolysis modeling, albeit with less-than-optimal codes that have been jerry-rigged from our reactor modeling work. Nevertheless, if the fields are as high as Bob indicates, the eventual production of nitric acid, in addition to the elevation of the redox potential, becomes a critical issue.	PK	This is a very complex phenomenon and bit is probably fair to conclude that the underlying theory has not been developed to the extent that "first principles" prediction is possble.
T3-11	H	Quiescent settling occurs when the flow velocity is so low and hence the hydrodynamic forces on particles are insufficient to prevent precipitation and settling.	PK	The basic physics of the process are known, but it appears that the models are insufficiently well-developed to provide quantitative prediction of this phenomenon.

T3-12	H	Electrostatic agglomeration occurs when the surfaces of different particles acquire surface charges of different sign, due to the pH of the solution lying between the PZCs (pH of zero charge) of the two surfaces.	K	The process is well understood at both the theoretical and practical levels and extensive use of this phenomenon is made in industry (e.g., electrophoretic painting of automobiles). Extensive PZC data are available for a wide range of oxides, oxyhydroxides, and hydroxides, and data are readily measured for those phases that have not been previously been studied.
T3-13	H	See T3-12. The surfaces may be modified chemically to change the surface charge characteristics and hence modify the agglomeration characteristics.	K	Modification of the surface charges is a commonly-practiced technology in industry and in the laboratory, so that there exists an extensive database exists on this subject. .
T3-14	H	See T3-12.	K	See T3-12.
T3-15	H	A wide variety of nucleation sites may exist in the debris-filled containment pool after a LOCA, and these may act as sites for the precipitation of various phases.	PK	The rate of nucleation of a phase onto a nucleus is determined by the Gibbs energy of nucleation. If this quantity is small, nucleation occurs readily, but if it is large nucleation is difficult and precipitation may not occur at all. The general principle is that "like nucleates like"; so that a carbonate nucleus will tend to nucleate carbonates, for example.

T3-16	H	Because radioactive material is expected to be trapped in the debris bed, the radiolysis of water is expected to exert a significant impact on the chemistry of the local environment. The principal impact will probably be due to the generation of hydrogen peroxide, which is a strong oxidizing agent that will oxidize Fe^{2+} to Fe^{3+}, for example.	K	The radiolysis of water has been modeled extensively in nuclear reactor technology to estimate the concentrations of the principal radiolysis products. These concentrations can then be used in Mixed Potential Models to estimate the redox potential and the corrosion potentials of various metals and alloys. The impact of radioactive species trapped in the debris bed can therefore be modeled quantitatively.
T3-17	M	If radioactive elements are concentrated at the sump screen, it is possible that the local redox conditions are such that high concentrations of certain species may form at that location.	PK	The redox potential, which is established by the radiolysis of water, is readily calculated and the impact of the redox potential on the solution chemistry may be assessed using various computer codes that are commercially available. In this way, the concentrations of the components in the system can be calculated and the possibility of highly localized concentrations can be assessed.

T3-18	M	The chemical environment within the sump screen is expected to significantly modify the morphology of fiberglass via deposition and etching, thereby altering the ability of the fiberglass to trap additional debris. Initial rankings changed due to following email, rankings indicated separately: Please find attached a table with my re-assessments of some originally "high" ratings of issues. On examining these issues again, it was evident that I did not fully recognize the impact of: 1. Time of exposure after the initial LOCA. 2. The amount of the specific material in the containment (e.g., zinc). My re-assessments also reflect the results of continuing radiolysis modeling, albeit with less-than-optimal codes that have been jerry-rigged from our reactor modeling work. Nevertheless, if the fields are as high as Bob indicates, the eventual production of nitric acid, in addition to the elevation of the redox potential, becomes a critical issue.	PK	The basic principles of deposition and etching are known (PZC, etc.), but the data for specific glasses do not appear to be currently available.
T3-19	M	Abrasion of pump seals could contribute to the debris inventory, but it would appear to be a relatively minor source in comparison with others.	PK	The basic principles of abrasion and wear ("fretting corrosion") are well-known, but the wear rate is a sensitive function of the substrate material. To this Reviewer's knowledge, the specific materials of interest have not been studied in this regard.

T3-20	H	Precipitation of material in heat exchangers, because of temperature gradients. The precipitated material could degrade the hydrodynamic and thermal performance of the exchanger.	K	The driving force for precipitation is the change in solubility, which can be effectively modeled using equil brium thermodynamics. This topic is well-developed in the oil and gas industry in relationship to the production of hydrocarbons.
T3-21	H	See T3-20	K	See T3-20
T3-22	H	Deposition of solids on the fuel, due to retrograde solubility, will degrade the ability of heat to be dissipated from the core and hence will lead to enhanced fuel temperatures.	K	The driving force for precipitation is the change in solubility, which can be effectively modeled using equil brium thermodynamics. This topic is well-developed in the oil and gas industry in relationship to the production of hydrocarbons.
T3-23	H	The reaction of Zr-based cladding with water will produce hydrogen and hence will contribute to the inventory of this gas in containment.	K	The reaction of zirconium alloys with water at high temperatures has been well-characterized (as a result of Three Mile Island). The impact of hydrogen on the redox potential of the pool environment and upon the reaction of this gas with oxygen are well-known and can be accurately predicted.
T3-24	H	The mixing of debris with water will reduce the effective heat capacity and hence will degrade the ability of the coolant to remove heat from the reactor core.	K	The heat capacity of a mixture can be accurately predicted by a mass-based mixing rule.
T3-25	H	Deposition of debris in coolant channels within the core may degrade the ability to cool the core in the event of a LOCA.	PK	Unl ke the deposition of a solid from solution, the prediction of blocking of channels by solid debris is much less certain, because of the complexity of the phenomenon.

T3-26	H	Settling of particulate debris in the core in low flow areas could result in reduction of heat transfer from the lower regions of the reactor fuel.	PK	While the physics of settling are understood, the application of the principles to reactors under LOCA conditions appears not to have been made.
T3-27	H	See T3-22	K	See T3-22
T3-28	L	Dissolution of uncoated concrete may result in a pH increase.	K	The pH of the containment pool is effectively buffered by the coolant boric acid, the TSP, and by the STB on the containment floor. It is difficult to imagine that the dissolution of concrete over the time available could significantly affect the pH of the containment pool.
T3-29	M	Dissolution of organic coatings (epoxy, alkyd, and Zn-base primers) may contribute to the debris inventory.	PK	Dissolution products will affect pool chemistry, including contr buting to chloride ion inventory, but the impact that these phenomena will have on the containment pool are not well-defined.

T3-30	M	Concentrated boric acid is known to be highly corrosive toward ferrous alloys, as evidenced by the Davis Besse incident. Initial rankings changed due to following email, rankings indicated separately: Please find attached a table with my re-assessments of some originally "high" ratings of issues. On examining these issues again, it was evident that I did not fully recognize the impact of: 1. Time of exposure after the initial LOCA. 2. The amount of the specific material in the containment (e.g., zinc). My re-assessments also reflect the results of continuing radiolysis modeling, albeit with less-than-optimal codes that have been jerry-rigged from our reactor modeling work. Nevertheless, if the fields are as high as Bob indicates, the eventual production of nitric acid, in addition to the elevation of the redox potential, becomes a critical issue.	PK	The potential for damage caused by concentrated boric acid is clear, but few data are available to quantify the effect.
T3-31	L	Radiolysis of carbonate solutions could result in the formation of organic acids.	K	While the radiolytic formation of organic acids may be possible under highly reducing conditions, in the presence of CO_2, the reverse is most commonly observed – the destruction of organic materials to produce CO_2.

APPENDIX H

PIRT Panelist Evaluations of T4 Phenomena:

ECCS Recirculation from 24 hours until 15 Days into LOCA

Table H-1: ECCS Recirculation (24 hours to 15 days) [a,b]

Issue Number	Phenomena	Description	Implications
T4-1	Source terms: Unsubmerged materials	Corrosion of various materials (Al, Fe, concrete, Cu, Zn) due to condensation and transport into containment pool; condensation expected due to non-uniform containment temperature	1. Contributes species to containment pool 2. Condensate chemistry & environment governs corrosion/leaching rates.
T4-2	Submerged source terms: Pb shielding	Any acetates present in containment pool will dissolve Pb, which could lead to formation of lead carbonate particulate. Lead blanketing to shield hot spots and covered with plastic coating, but coating likely destroyed. Several hundred pounds of lead in flat sheets. Some plants still use lead wool. No Pb in ICET program.	Would provide additional particulate loading within containment pool
T4-3	Submerged source terms: Cu	Concern stems not from Cu compounds, but the various effects that Cu may have on other corrosion processes. Cu concentrations evaluated in ICET program; Cu comes from containment air coolers, motor windings and grounding straps.	1. By forming a galvanic couple, can facilitate attack of other metals (e.g., Al). 2. Cu ion deposition can occur which may inhibit corrosion. 3. Within an oxygenated environment, Cu can accelerate corrosion
T4-4	Submerged source terms: Fe	Boron inhibits pitting corrosion on surface of steel structures. Fe concentrations evaluated in ICET program.	1. Less Fe ions in containment pool. 2. less corrosion and structural weakening of steel structures
T4-5	Submerged source terms: Al, decreased concentrations	Less dissolved Al, than in ICET #1, affects the type and quantity of chemical byproducts that form at high pH.	1. Less corrosion inhibition of insulation results in greater Si levels. 2. Less Al corrosion products allows other species to form within pool
T4-6	Submerged source terms: Al*, increased concentrations	Specific mix of other metals in containment pool leads to more Al corrosion (catalytic effect) at either lower or higher pH	Increases Al solid species which may form.

T4-7	Submerged source terms: Fiberglass dissolution	Dissolution at high pH contributes Silicates, Cu, Na, Al, Mg, B, misc. organics to containment pool. F berglass concentrations in ICET test plan	1. Solid species formation is affected by silicates, Cu, Na, Al, Mg, B concentrations. 2. Misc. organics in sump pool affect complexation.
T4-8	Submerged source terms: Fiberglass inhibition	Leaching of fiberglass inhibits corrosion of certain metals (e.g., Al)	Less metallic ions within containment pool
T4-9	Submerged source terms: Zn passivation	Zn passivation decreases corrosion passivation of Fe which allows Fe corrosion.	Galvanized steal/zinc coating; vessel site for ...; pH>8, dissolve ...; precipitate as silicate; ZnO, ZnOH corrosion products;
T4-10	Submerged source terms: Zn corrosion products	Zn oxide and hydroxide corrosion products form which contributes to solid species loading.	Additional source term that could contr bute to sump screen clogging and/or downstream effects.
T4-11	Submerged source terms: Zn coprecipitation	Zn may coprecipitate with other species (, notably Fe, Al). Reactor Vessel a poss ble site for coprecipitation	Additional source term that could contr bute to sump screen clogging and/or downstream effects.
T4-12	Zinc hydroxide dissolution (zincate)	Buffered pH containment pool dissolves zinc hydroxide so that other species, particularly silicates, may precipitate	Zinc silicates may be more likely to contribute (more transportable, possibly amorphous) to head loss or downstream effects.
T4-13	Submerged source terms: Zn-based coatings	Leaching of Zn-based primers (primarily inorganic Zn phosphates) creates possibility for additional dissolved Zn in solution	Lead to formation of zinc oxide/hydroxides, carbonates, or silicates that could contribute to head loss/downstream effects.
T4-14	Source Term: Seal table corrosion	Multiple materials in seal table (Inconel, SS, plastics, organics) leads to formation of additional dissolved species. Seal table is an area in bottom of core where instruments are inserted into the core. A series of pressure isolation chambers. Generally lower than bottom of containment floor	Additional dissolved species present within the containment pool

T4-15	Submerged source terms: Fire barriers	Silicon-based seals (bisco seals) used at many structural penetration points at fire barriers. These may leach in post-LOCA environment	Another silicon source term that could contribute to formation of chemical byproducts which could impact head loss and downstream performance.
T4-16	Submerged source terms: organic buoyancy	Organics coat materials (e.g., aluminum, cal-sil, fiberglass, etc.) and increase the buoyancy of particulate so that it's more likely to float	This could enhance transport of particulates or chemical byproducts to the sump screen.
T4-17	Submerged source terms: coatings	Leaching of submerged coatings to contribute species to the containment pool. Possible sources include Pb-based paints (older containment buildings), phenolics, PVC	Additional source terms (especially chlorides, fluorides) which contribute to containment pool chemistry.
T4-18	Submerged source terms: RCP oil tank failure	Overflow, failure, or leakage occurs either by LOCA or preexisting condition. Oils and other organics are released into containment pool.	Oils and other organics may affect complexation and sequestration of metallic species. Containment pool chemistry and byproduct production are influenced.
T4-19	Submerged source terms: biofilm formation	Biofilms form which protect against metallic corrosion (by forming passive layer) or lead to production of acids which increase metallic corrosion. Some biofilms can be preexisting in containment from outages.	1. If decreases metallic corrosion, would reduce containment pool concentration of certain dissolved species. 2. If increases metallic corrosion, would increase containment pool concentration of certain dissolved species. 3. May eventually lead to SCC failures of components.
T4-20	Submerged source terms: biologically enhanced corrosion	General corrosion is enhanced due to biological agents. Examples include polysaccharides and sulfate reducing bacteria which may enhance Fe corrosion. Sulfur or carbon sources contribute to bacteria formation.	Fe and other metallic corrosion is enhanced leading to increased concentrations of metallic species within the containment pool.
T4-21	Submerged source terms: biologically enhanced H_2 embrittlement	Anaerobic and aerobic bacteria synergistically enhance stress corrosion cracking and hydrogen embrittlement of steels. Sulfur or carbon sources contribute to bacteria formation.	Structural cracking enhanced which could weaken and lead to failure of structural components.

T4-22	Submerged source terms: biological growth in debris beds	Bacteria grow in preexisting debris beds located on the sump strainer screen or elsewhere within the ECCS system.	Increased source term which can contr bute to clogging or detrimental performance of pumps, valves, etc.
T4-23	Submerged source terms: electrical insulation	Chlorides form due to radiological breakdown of cable insulation. This affect was simulated in the ICET experiment.	1. Additional source term for products affecting the containment pool chemistry. 2. Chlorides affect radiolysis and ionic strength of containment pool environment.
T4-24	Reactor Core: Fuel spalling	Zr-oxide layer spalls off reactor fuel due to temp gradient - 50 micron layers exist	1. Heat transfer depends on balance of oxide growth, deposition, and particulates, particulates may also enhance coagulation in core that may diminish heat transfer of water. 2. Additional solid products which contr bute to clogging within the reactor core. 3. Additional products which could lead to clogging within the reactor core or be transported back to the sump screen.
T4-25	Radiological** effects: debris bed accumulation	Radio nuclides transport to, accumulate, and become concentrated within the sump screen debris bed altering the local chemical conditions	1. More concentrated effect than if spread throughout containment pool 2. Directly alter chemical byproducts formed at the sump screen, contributing to head loss.
T4-26	Radiological** effects: Dissolution & oxidation changes	Radiolysis leads to changes in the oxidation state in reactor, piping, and containment materials which affects the dissolution rate and the byproducts which form.	Affects the chemical byproduct source term loading that is present in the containment pool.
T4-27	Radiological** effects: Radiolytic affect on biofilms	Radiolysis inhibits the formation and growth of biofilms	Films may be either advantageous or detrimental to corrosion and chemical byproduct formation (see earlier items)

T4-28	Radiological** effects: Redox potential changes	Radiolysis causes changes in the redox potential either in the containment pool or in debris beds (on the sump screen, or within ECCS component)	Could affect chemistry in containment pool which affects species which form (e.g., Hanford tank: mixture of different components @ different phases)
T4-29	Radiological** effects: corrosion rate changes	Low doses and low temperature elevate the corrosion rate through formation of hypochlorite through radiolysis of Cl bearing water. Could increase corrosion rates (esp. pitting corrosion) of Al, SS, Fe.	Corrosion rate increases could increase the amount of metallic and nonmetallic species in containment; alters chemical byproduct formation.
T4-30	Radiological** effects: agglomeration	Radiolysis enhances agglomeration of chemical species.	Advantage: Agglomeration could lead to increase settling rate of product. Disadvantage: Could form products that are more likely to cause clogging of sump screens, downstream components.
T4-31	CO_2/O_2 air exchange	Air exchange provides source for CO_2 ingestion within the containment pool. CO_2 quantities limited by containment volume. Radiolysis can also promote formation of carbonates.	Increase solid species in containment pool.
T4-32	ECCS Pumps: erosion/corrosion	Chemical byproducts cause erosion or corrosion of pump internals, especially tight tolerance components (bearings, wear rings, impellers, etc.).	Pump performance degrades, possibly to the point of being inoperable.
T4-33	ECCS Pumps: seal degradation	Chemical environment causes leaching/degradation of pump seal materials or chemical byproducts cause seal erosion	Pump performance degrades, possibly to the point of being inoperable.
T4-34	Heat exchanger: secondary contaminants	Leaks in heat exchanger tubes allow secondary-side water additives to migrate over to the containment pool.	Additives in the secondary side water chemistry alter the chemistry in the containment pool.

T4-35	Heat exchanger: precipitate formation	Drop in temperature leads to the formation of solid species (e.g., AlOOH, FeOOH amorphous SiO_2) and ripening such that macroscale coatings and/or suspended particulates form.	1. Create clogging, and less efficient heat transfer for cooling the ECCS recirculating water. 2. Form products that may clog reactor core and degrade heat transfer from fuel. 3. Products form which transport back to sump screen and lead to head loss. 4. Particulates act as nucleation sites for other compounds to precipitate
T4-36	pH drop in containment pool	pH drop due to evolving containment pool chemistry causes Al (and other metallic species) to precipitate.	Creates solid particulate loading that could affect sump screen head loss, heat exchanger clogging, reactor core clogging, and additional nucleation sites.
T4-37	Reactor Core: continued deposition/precipitation	Zn, Ca, Mg, CO_2 based deposits, films, and precipitates may form at higher temperatures within the reactor core	1. Diminished heat transfer from the reactor fuel. 2. Spalling could create additional solid products which contribute to clogging within the reactor core. 3. Spalling could create additional solid products which contribute to sump screen head loss.
T4-38	Reactor Core: fuel deposition spalling	Zn, Ca, Mg, CO_2 based deposits and films which form on the reactor core spall	1. Additional solid products which contribute to clogging within the reactor core. 2. Additional solid products which contribute to sump screen head loss.
T4-39	Transport Phenomena: amorphous coating	Amorphous silica forms on surface of chemical precipitates (e.g., AlOOH, FeOOH). Coatings increase density and make it less likely for products to transport to sump screen.	Decreased transportability will result in less product migrating to or through the sump screen.
T4-40	Transport Phenomena: precipitation/co-precipitation	Precipitation/co-precipitation and ripening of solid species within containment pool creates solid species which are less likely to transport.	Decreased transportability will result in less product migrating to or through the sump screen.
T4-41	Transport Phenomena: Metallic scouring	Turbulent flow causes chemical films/products to be scour off the surface of metallic components so that they enter the containment pool.	Increased particulate loading that may increase head loss, degrade pump, valve, heat exchanger performance, and/or affect heat transfer from reactor fuel.

[a] Sump pool temperatures are expected to range between 50 – 70C during this time period.

[b] Phenomena listed in earlier time periods (i.e., T1, T2, T3) may continue to be active or important during this time.
[c] Al, in ICET tests, has provided an important contribution to observed chemical effects both through precipitate formation at high concentrations and pH and in inhibiting fberglass leaching.
[d] Soluble/insoluble activation species include, for example, Na^{24}, Be^{7}, Cr^{51}, $Co^{58,60}$, $Mn^{54,56}$, $Zr^{95,97}$, $Fe^{55,59}$, $Ni^{59,63}$, $Nb^{95,97}$, Sb^{125}, Zn^{65}, $Cs^{134,137}$, and $I^{131,133}$

Table H-2: Apps T4 Ranking Summary

Issue Number	Importance (H, M, L)	Rationale	Knowledge (K, PK, UK)	Rationale
T4-1	L	Effect relatively small compared with submerged material.	PK	ICET series provide some information. More is needed to evaluate non-isothermal effects.
T4-2	L	Amount of exposed lead not expected to be large. Coatings of Pb hydroxy-carbonates expected to form. Low concentrations of acetate expected in recirculating coolant are unlikely to cause significant dissolution. Possible passivation through formation of complex Pb silicates.	PK	If cooling water composition can be estimated, its impact on lead dissolution could be calculated.
T4-3	L/M	Copper does not corrode rapidly. Dissolved Cu would plate out on Fe and Al, but amount available in solution would be small.	UK	Effect of precipitated Cu on galvanic corrosion of other metals requires quantification.
T4-4	L	Fe corrosion would be slow.	PK	ICET series and evaluation at CNWRA provide some information.
T4-5	L	Decreased Al corrosion would mitigate sump screen blockage.	PK	ICET series and corrosion studies at CNWRA provide some information.
T4-6	H	Al corrosion could strongly affect sump screen blockage.	PK	ICET series and corrosion studies at CNWRA provide some information.
T4-7	M	Dissolution of fiberglass only moderate.	K	ICET series and dissolution studies at CNWRA provide information.

T4-8	M	Effect of fiberglass dissolution on Al corrosion would be variable.	PK	ICET series and dissolution studies at CNWRA provide some information.
T4-9	L	Overall impact would be small	PK	ICET series and evaluation at CNWRA provide some information.
T4-10	L	Total Zn dissolution would be small.	PK	ICET series and evaluation at CNWRA provide some information.
T4-11	L	Total Zn dissolution would be small.	PK	ICET series and evaluation at CNWRA provide some information.
T4-12	L	Total Zn dissolution would be small.	PK	ICET series and evaluation at CNWRA provide some information.
T4-13	L	Zn based primers are relatively insoluble.	PK	ICET series provide some information. Additional calculations to evaluate primer solubility are needed.
T4-14	L	Materials are resistant to corrosion, and corrosion products would be small in amount.	PK	Additional calculations to evaluate alloy corrosion are needed. Data may be available from manufacturers.
T4-15	L	Minor source of contamination.	UK	No information is available on the leaching characteristics.
T4-16	M/H	Flotation characteristics are important in relation to sump screen blockage.	PK	Some information available from head loss testing.
T4-17	L	Effect on coolant composition and formation of secondary precipitates is likely to be small.	UK	Further review of PWR-specific design is required.

T4-18	M	Oil leakage could cause flotation of debris and impede metal corrosion.	UK	Nature of contaminant release and physical and chemical effects require evaluation.
T4-19	L	Environment is likely to be relatively free of biofilms.	UK	Would require plant specific characterization.
T4-20	L	Time frame and temperature would limit extent of biological reactions.	UK	Would require plant specific characterization.
T4-21	L	Time frame and temperature would limit extent of biological reactions.	UK	Would require plant specific characterization.
T4-22	L	Time frame and temperature would limit extent of biological reactions.	UK	Would require plant specific characterization.
T4-23	L	Duration is too short for a major effect.	UK	Extent of radiolysis would depend on plant specific conditions.
T4-24	L/M	Spallation of Zr oxides would add only a small burden to the particulate load, but could contribute to core damage through restriction of circulation.	UK	Actual behavior would require laboratory and pilot testing.
T4-25	L/M	Concentration of radionuclides in the bed may not have a major impact in altering the chemistry of the system except perhaps with respect to organics.	UK	Specific experiments are needed to quantify the effect.
T4-26	L	Overall effect expected to be small. Oxidation state would already be high.	UK	Need to evaluate the effect due to radiolysis.

T4-27	L	Problem not likely to have a major effect during the period in question. Inhibition effects would likely be small.	UK	Effect would have to be evaluated.
T4-28	L/M	Radiolysis would likely enhance the oxidation state, but conditions would already be oxidizing. There could be an acceleration of oxidation reactions.	PK	Effect of radiolysis on the oxidation state can be estimated, but plant specific conditions would be needed to make an evaluation.
T4-29	L/M	Effect could accelerate Al corrosion. Effect likely to be modest as concentration of Cl expected to be relatively low.	PK	Effect could be estimated, but experiments are needed to quantify rates.
T4-30	L/M	Effect could be important, but is difficult to quantify.	UK	Little is known on the subject.
T4-31	L	Effect likely to be small.	PK	It might be possible to calculate the magnitude of this effect.
T4-32	M	Effect could be serious, but surely pumps are designed to prevent such an outcome?	PK	Manufacturers should be consulted.
T4-33	M	Effect likely to be small.	UK	Evaluation needed. Manufacturers should be consulted.
T4-34	L	Effect likely to be small.	UK	Effect of contaminants would have to be evaluated.
T4-35	M/H	Could adversely affect heat exchanger operation and cause costly repairs.	PK	Likelihood of precipitation and type could be evaluated

T4-36	L	pH drop in buffered system is likely to be small.	PK	Could be estimated.
T4-37	M/H	Could adversely cooling and core damage could occur, with costly repairs.	PK	Likelihood of precipitation and type could be evaluated.
T4-38	M	Spallation would partially restore cooling ability. Spallation products not likely to constitute a large volume of material.	UK	Difficult to estimate spallation. Testing would be required.
T4-39	L/M	Amount of potential precipitate removed by precipitation on surfaces is likely to be small in relation to material precipitating in suspension. However, effect is dependent on pH buffer system used.	PK	ICET series could provide some information. Specific testing would be required.
T4-40	M	Phenomenon could limit screen blockage, but is dependent on plant design and choice of pH buffer.	UK	Effect would depend on plant specific conditions.
T4-41	L/M	Corrosion rates could be greatly accelerated, especially with respect to Al.	UK	Effect is difficult to quantify. Much depends on the location of the break and proximity of exposed metal structures.

Table H-3: Chen T4 Ranking Summary

Issue Number	Importance (H, M, L)	Rationale	Knowledge (K, PK, UK)	Rationale
T4-1	L		PK	
T4-2	H		PK	
T4-3	M		PK	
T4-4	H		PK	
T4-5	L	I changed initial ranking from high to low based on email received from Chen	K	
T4-6	H		K	
T4-7	L		PK	
T4-8	L		PK	

T4-9	L		PK	
T4-10	M		PK	
T4-11	L		PK	
T4-12	M		PK	
T4-13	M		PK	
T4-14	L		PK	
T4-15	L		UK	
T4-16	L		UK	
T4-17	L		UK	
T4-18	M		PK	

T4-19	L		UK	
T4-20	L		PK	
T4-21	L		UK	
T4-22	L		UK	
T4-23	M		PK	
T4-24	L		PK	
T4-25	L	too low a residence around the sump screen	UK	
T4-26	L		UK	
T4-27	L		PK	
T4-28	M		PK	

T4-29	M		PK	
T4-30	L		UK	
T4-31	L		PK	
T4-32	M		K	
T4-33	M		K	
T4-34	L		PK	
T4-35	L		K	
T4-36	M	This ranking was changed based on following email: "As discussed today, the [containment pool] pH could change between 9 and 4.5 which is significant to warrant a High [ranking]. As the chance of getting to pH=4.5 is little, I can live with a rating of Medium but not less."	PK	
T4-37	L		PK	

T4-38	L		UK	
T4-39	H	amorphous silica particles can enhance sump screen plugging	PK	
T4-40	M		PK	
T4-41	M		PK	

Table H-4: Delegard T4 Ranking Summary

Issue Number	Importance (H, M, L)	Rationale	Knowledge (K, PK, UK)	Rationale
T4-1	L	Little condensate flow and little corrosion observed in ICET.	K	ICET showed little corrosion attack in condensing region.
T4-2	L	Anticipate low lead corrosion rate in the system.	PK	Some scoping tests would aid in determining severity of this phenomenon.[1]
T4-3	M	The effect of galvanic couples was not examined but could be strong, especially for aluminum.	UK	See T2-22. [1]
T4-4	L	ICET and other testing showed little iron corrosion. However, effects of high and low temperature surfaces need to be studied (see T3-20 to -26).	PK	Studied in ICET and other testing. [1]
T4-5	M	Though relatively important, this situation has already been examined in ICET and other testing.	PK	Studied in ICET and other testing. [1]
T4-6	M	Though relatively important, this situation has already been examined in ICET and other testing.	PK	Studied in ICET and other testing. [1]
T4-7	M	Though relatively important, this situation has already been examined in ICET and other testing.	PK	Studied in ICET and other testing. [1]
T4-8	M	Though relatively important, this situation has already been examined in ICET and other testing.	PK	Studied in ICET and other testing. [1]

T4-9	M	Though relatively important, this situation has already been examined in ICET and other testing.	PK	Studied in ICET and other testing. [1]
T4-10	M	Though relatively important, this situation has already been examined in ICET and other testing.	PK	Studied in ICET and other testing. [1]
T4-11	M	Though relatively important, this situation has already been examined in ICET and other testing.	PK	Studied in ICET and other testing. [1]
T4-12	M	Though relatively important, this situation has already been examined in ICET and other testing.	PK	Studied in ICET and other testing. [1]
T4-13	M	Though relatively important, this situation has already been examined in ICET and other testing.	PK	Studied in ICET and other testing. [1]
T4-14	L	The materials in the seal table appear to be similar to materials in other areas exposed to the coolant. The seal table area also seems to be in a relatively stagnant location and, being removed from the flow currents, might not contribute significantly to the solute or suspended solids burden.	PK	Assessment by reactor engineer needed to validate bases of judgment that seal table materials influence is of low importance.
T4-15	M	The effects of thermolysis and radiolysis on the bisco seals should be assessed with those of the organic materials (paints, plastics).	UK	See T2-24 and T3-29.
T4-16	H	The flotation phenomenon has the potential to motivate a lot of solids to the sump screen.	UK	This mechanism looks like froth flotation used in ore beneficiation.

T4-17	H	Similar to T2-24 and T3-29.	UK	See T2-24 and T3-29.
T4-18	M	See T1-7. Effects of oil not studied; oil may accrete small solids and drift downstream to span matted fibers at sump screen.	PK	Effects of oil to accrete finely particulate solids under this system not known but example exists in oil spills from oil tankers onto beaches.
T4-19	?	Any consideration of biofilms must first evaluate the potential for biota to survive in ~0.2 molar boron. Note that borates are used as algaecides.	UK	The essentially steady presence of ~2000 ppm boron (about 0.2 moles per liter) is ensured by the injection of the emergency core cooling water about 20-30 minutes post-LOCA.
T4-20	?		UK	
T4-21	?		UK	
T4-22	?		UK	
T4-23	L	The ~100 ppm chloride seemingly had little effect on the solids loading. However, other radiolytic effects on the electrical insulation may have larger effects on the solids behavior (see T2-24, T3-5, and T3-29, for example).	K	Tested in ICET and other experiments.
T4-24	M	This is addressed in T1-1.	PK	Crud releases observed during reactor shut-down and spent fuel storage.
T4-25	M	This is addressed in T3-16.	UK	Testing of the effects of high radiation fields needed for sump screen beds. Effects expected to be higher for the organic materials.

T4-26	M	The effects of radiolysis on the inorganic components are expected to be lower than those of the organic materials.	UK	Testing is necessary as suggested in T2-20 and T3-6.
T4-27	?	See T4-19 through T4-22 for speculation on the detrimental effects of borate on biota. Radiolysis would decrease the potential for biofilms further.	UK	The essentially steady presence of ~2000 ppm boron (about 0.2 moles per liter) is ensured by the injection of the emergency core cooling water about 20-30 minutes post-LOCA.
T4-28	M	Similar to T4-26.	UK	Testing is necessary.
T4-29	M	Oxidizing conditions imposed by high dose radiolysis of chloride brines to form hypochlorite; low chloride in ECCS may not be sufficient to form hypochlorite.	PK	Tests of aluminum in waters containing 100 ppm chloride and 100 ppm H_2O_2 showed no enhanced corrosion at ~92°C.
T4-30	M	Agglomeration could enhance sump screen blockage	UK	Testing is necessary.
T4-31	M	See T2-8.	K	See T2-8.
T4-32	H	For pump seal and erosion issues, see T3-19.	UK	See T3-19.
T4-33	H		UK	

T4-34	L	Contributions of solutes from secondary cooling likely to be very low. Secondary coolant leakage only can dilute solids and solutes concentrations.	K	If anything, serves to dilute solids and solutes concentrations.
T4-35	H	See T3-20.	UK	See T3-20.
T4-36	L	The quantity of dissolved aluminum is already low at pH 10; pH decrease will have little additional effect. Differences in metal solubility represent little difference in the total solids loading in the suspended solids budget.	K	Aluminum, zinc, and lead are the only amphoteric metals in the system (iron, zirconium, copper are not amphoteric).
T4-37	H	See T3-27.	UK	See T3-27.
T4-38	H		UK	
T4-39	H	The continued dissolution of Cal-Sil followed by the precipitation of amorphous silica will occur. The precipitating silica can span over the solids bed on the sump screen.	PK	This effect is amenable to investigation through the parametric studies suggested for T4-3 to T4-13
T4-40	H		PK	

| T4-41 | L | The incremental addition of solids scoured from surfaces after the first day is expected to be small unless large, dense, and abrasive pieces (e.g., UO_2 fuel bits) are added to the system. Most solids found in ICET were finely particulate (e.g., aluminum hydroxide, tobermorite) and/or softer than metal (e.g., fiberglass). | PK | Corrosion/erosion testing may help confirm the effects of scouring. |

[1] Note that the parametric small-scale testing may provide a better overall assessment of the chemical interactions suggested in phenomena T4-2 through T4-13. The testing would study the effects, on the solids listed in phenomena T4-2 through T4-13 plus Cal-Sil and concrete, of solution composition including pH and dissolved phosphate and borate. Cal-Sil is a special case in that it contributes significant calcium and silicate to the solution and also affects the pH. The testing would examine the formation of solid phases and the rates of material corrosion in the various systems.

Table H-5: Litman T4 Ranking Summary

Issue Number	Importance (H, M, L)	Rationale	Knowledge (K, PK, UK)	Rationale
T4-1	L	Subsequent to spray termination the contribution to recirculation water dissolved/precipitated materials from unsubmerged components will be minimal. It will be limited by the rate of condensation that should be a small factor during this time period.	PK	Without the presence of containment spray there is no driving force to remove precipitated materials from unsubmerged surfaces.
T4-2	M	Most lead shielding n containment will be covered with plastic coatings. The overall dissolution of lead should only result in concentrations on the order of ppb to small ppm range. However, lead is known to induce cracking in stainless steel components.	PK	This is a long term effect which is minimized by the low potential concentration of lead and the neutral pH conditions..
T4-3	M	All containment-building components are grounded using either copper or aluminum grounding straps. These are generally uncoated and found at the bottom of containment; thus they will be submerged.	PK	Precipitated materials may actually be 'attracted' to these grounding materials coating them, and inhibiting the dissolution of copper. However, this has not been effectively tested. Some ICET results indicate that this is potentially true.
T4-4	H	The inhbition of carbon steel corrosion by boric acid occurs in combination with other corrosion control agents under specific concentration ranges. The combination of soluble and insoluble ionic materials in the recirculated water will be more suited towards corrosion rather than inhibition.	K	PWR operating history with carbon steel and mild steel corrosion with boric acid.

T4-5	L	During this time period the water level in containment should remain fairly constant. Equilibrium between dissolved and precipitated species will be reached.	PK	Results of ICET indicate that within about 10-20 days the equilibrium condition would be reached for dissolution of materials.
T4-6	L	During this time period the water level in containment should remain fairly constant. Equilibrium between dissolved and precipitated species will be reached.	PK	Results of ICET indicate that within about 10-20 days the equilibrium condition would be reached for dissolution of materials.
T4-7	M	Equilibrium conditions should be established. The maximum amount of materials to be leached ant precipitated should have already occurred by day 15.	PK	Results of ICET programs demonstrate the equilibrium is eventually established.
T4-8	M	This particular material may continue to contribute in the longer term due to radiolytic effects on f berglass insulation. This is not true with other species.	K	Insoluble silicates, silicon dioxide and other silicon compounds are subject to radiolytic dissolution (PWR spent fuel pool silica issues).
T4-9	L	After several days settling of precipitates will most likely coat many metallic surfaces causing them to react more slowly with the bu k water.	PK	Results of ICET programs with these material coupons show minimum dissolution after initial periods of exposure.
T4-10	L	After several days settling of precipitates will most likely coat many metallic surfaces causing them to react more slowly with the bu k water.	PK	Results of ICET programs with these material coupons show minimum dissolution after initial periods of exposure.
T4-11	L	This effect would have already occurred.	PK	ICET coupon tests.
T4-12	L	After several days settling of precipitates will most likely coat many metallic surfaces causing them to react more slowly with the bu k water.	PK	Results of ICET programs with these material coupons show minimum dissolution after initial periods of exposure.

T4-13	L	Amount of coating is limited. Amount submerged my be a significant portion or that available. Changed ranking from M/L to L based on following email: "Initially I denoted this as "L?". Digby had rated this as "H". Based on the time frame, the pH of the solution and the limit on the amount of zinc that can be in containment, this should definitively be an L. The zinc that dissolves or leaches from the metallic surface or primers would almost immediately be precipitated as the hydroxide. In the quiescent flow of the containment this material would form a protective barrier, inhibiting additional dissolution of zinc."	UK	Zinc based coatings have not been tested in this environment. It is unknown what effect radiolysis will have.
T4-14	L	The seal table is in a low flow area, and most likely sill not significantly contribute to dissolved materials. May be a source of non-metallics.	UK	Each plant will have varying designs for this part of plant. Would need more information for all plants to determine long range effects.
T4-15	L	These fire barrier materials are also significantly 'waterproof'. They are organic polymers formed using a metallic (proprietary) catalyst.	UK	Specific testing would have to be performed to ascertain the contribution from these materials in the post-LOCA environment.
T4-16	L	Materials of low enough density to float would most likely be ground up in pump impellers without causing significant pump damage.	PK	These materials would generally be soft and not contribute to component wear.
T4-17	M	Only if various metallic based coatings are present	PK	Plant design dependent.

ID		Rationale		Comments
T4-18	M	Any small leaks of pump oils or raw water systems in containment will provide additional materials for precipitation and bacterial growth.	PK	Several plants have raw water systems in containment. All plants have RCP oil and other small pump oil reservoirs that could add to organic loading. Long term corrosion of these components could lead to loss of materials from these systems.
T4-19	M	Containment buildings contain bacteria and algae from exchange with outside air during normal operation. Raw water leaks and foot traffic prior to event will also contribute to bacterial/spore loading.	PK	PWR experience has generally shown that biofilms are detrimental to plant components. With no controls on biofilm growth, under deposit corrosion is very likely. This will lead to SCC and pitting corrosion.
T4-20	M	Containment buildings contain bacteria and algae from exchange with outside air during normal operation. Raw water leaks and foot traffic prior to event will also contribute to bacterial/spore loading.	PK	PWR experience has generally shown that biofilms are detrimental to plant components. With no controls on biofilm growth, under deposit corrosion is very likely. This will lead to SCC and pitting corrosion.
T4-21	M	Depends on the type of bacteria and the oxygen supply available in containment.	UK	It is not clear how long the oxygen supply in containment will last with large scale corrosion taking effect.

| T4-22 | H | Debris beds will provide a 'nesting' location for bacteria and spores. Any organic materials (food source) will l kely be trapped there.
Additional rationale received in email: Containment buildings are routinely exposed to outside air during refueling outages. The large refueling hatch is open and the containment building is operated under a negative pressure (this ensures no gaseous or particulate releases). Refueling outages are routinely scheduled for spring and fall, the two times of the year when there is the greatest amount of fungal and bacterial spores in the air, not to mention pollen. These materials will get sucked into the containment building. Under normal operating conditions these generally do not create a problem, but following a LOCA where debris gets concentrated, the food source for these microbes will be in close contact with them. The biological products from growth are polysaccharides that are sticky, tenacious materials. These compounds are known to cause fouling in other systems in power plants, especially on stainless steel surfaces. They initially lead to reduction in heat transfer capability, but eventually cause pitting and under deposit corrosion, leading to component failures.
Conditions in containment are rife for their evolution and production. Their growth starts almost immediately but left unchecked increases exponentially and within 15 days may create significant hallenges to both heat transfer and low.
I think that this should be rated 'high' in this time frame | PK c f | Industry experience in these areas is rare, however this has occurred in raw water systems on traveling screens. |
|---|---|---|---|---|

T4-23	L	The breakdown of cable insulating materials would slow down at this point	PK	Long term tests of the effects of radiation to the cable coatings in this liquid environment would need to be done to confirm this.
T4-24	M	The CRUD layer on the fuel surface will begin to spall as a result of changing fuel centerline temperature (loss of decay heat effect).	K	This effect can be easily observed in the SFP of nuclear power facilities with older fuel assemblies.
T4-25	M	As more debris accumulates the possibility to concentrate radionuclides by coprecipitation increases. Any observed radiolytic effects will also increase.	PK	Effects will depend on the material on the debris bed. Testing of specific materials would me necessary to determine the overall effect.
T4-26	L	During this time period, the general radiolytic effects should have reached equilibrium (except in those areas where radionuclides will accumulate see T4-25)	UK	Specific testing would need to be done.
T4-27	L	Bacteria and algae are much more resistant to the effects of radiolysis than other materials.	PK	Significant events at PWRs where oil spills in SFPs have lead to growths of biofilms and algae even in the presence of high exposure from spent fuel.
T4-28	M	These effects will reach equilibrium. The build up of hydrogen will continue but would be mitigated through the use of hydrogen recombiners.	PK	Plant design. Need tests on radiolytic effects to solidify this factor.
T4-29	L	The concentration of chlorides will generally be low. Formation of hypochlorite will occur but at low concentrations so as not to cause significant oxidizing effects.	PK	Limited chloride source term.
T4-30	M	The presence of ionizing radiation in solution will increase net 'charge' in solution favoring agglomeration of suspended materials.	PK	Flocculating agents work better at higher ionic strengths

T4-31	M	Air exchange would be minimized following a LOCA in order to reduce radioactive emissions. Thus a minimal amount of CO_2 would be introduced into containment.	K	With containment spray off and minimal make up water the source of carbon dioxide would most l kely be depleted with the first day.
T4-32	M	During this time period the continued presence of two phase flow through the pumps will have some effects on metallic erosion.	PK	These pumps are designed as "pure" water pumps.
T4-33	M	During this time period the continued presence of two phase flow through the pumps will have some effects on metallic erosion.	PK	These pumps are designed as "pure" water pumps.
T4-34	M	RHR heat exchangers are cooled wither by raw water or component cooling systems.	K	The cleanliness levels of the systems that may leak into the RCS is not the same as the RCS and containment. This will contribute contaminants that will have various effects.
T4-35	H	Precipitate formation due to temperature drops will also trap other soluble materials. Any particulates that normally would pass through the HX will also be trapped by precipitate.	K	Long-term operation of plant heat exchangers has shown that debris can bridge multiple tubes and lead to reduced flow. This is documented in PWR databases for heat exchanger maintenance.
T4-36	L	The changes in pH expected this long after the event are very small.	K	The pH function will be unaffected due to the large buffering capacity after the leaching of materials with the buffering agent.
T4-37	L/M	This can be mitigated by plant operators because reversal of flow can be performed on a routine basis.	K	Plant design characteristics allow for the change of flow from hot leg to cold leg based on core cooling needs.
T4-38	M	As the fuel continues to cool spallation from the surfaces will continue.	K	SFP bottom deposits confirm that this happens routinely as fuel cools down over months of storage.

T4-39	L	It is unlikely that amorphous SiO_2 will exist due to radiolytic effects.	K	Silica compounds in water are know to decompose to yield soluble silicates in the presence of a strong radiation field.
T4-40	M	As precipitates become denser their transportability will decrease.	K	Precipitates settle better after aging and increasing in density.
T4-41	L	The Low flow conditions that exist would not be significant ot cause general scouring.	PK	Plant specific design features.

Table H-6: MacDonald T4 Ranking Summary

Issue Number	Importance (H, M, L)	Rationale	Knowledge (K, PK, UK)	Rationale
T4-1	L	Corrosion of unsubmerged material due to condensation may contr bute to the transfer of material into the containment pool. Initial rankings changed due to following email, rankings indicated separately: Please find attached a table with my re-assessments of some originally "high" ratings of issues. On examining these issues again, it was evident that I did not fully recognize the impact of: 1. Time of exposure after the initial LOCA. 2. The amount of the specific material in the containment (e.g., zinc). My re-assessments also reflect the results of continuing radiolysis modeling, albeit with less-than-optimal codes that have been jerry-rigged from our reactor modeling work. Nevertheless, if the fields are as high as Bob indicates, the eventual production of nitric acid, in addition to the elevation of the redox potential, becomes a critical issue.	K	This is essentially an "atmospheric corrosion" problem, which has been extensively studied in corrosion science.
T4-2	H	Corrosion of submerged material due to condensation may contribute to the transfer of material into the containment pool.	K	The corrosion of materials has been extensively studied and the rate of production of corrosion products can be predicted, in many cases.

T4-3	M	Copper, which is present in containment, may corrode to produce corrosion products that will contribute significantly to the inventory of solid products.	UK	Copper corrodes only under oxidizing conditions. The environment that will exist post-LOCA is probably reducing in nature, due to the presence of hydrogen, but this issue must be resolved by modeling.
T4-4	L	The corrosion of iron is inhibited by borate ion and this will lead to a lower inventory of iron corrosion products than otherwise would be the case.	K	There is little evidence that borate ion significantly inh bits the general corrosion of iron; recent evidence suggests that it does inh bit localized corrosion (e.g., pitting).
T4-5	M	The corrosion of aluminum is inhibited by corrosion products and other material precipitated on the metal surface.	PK	The corrosion of Al is well understood and under some circumstances the corrosion rate is reduced by deposits. Prediction of the precise conditions is problematic.
T4-6	M	Possibility of the corrosion of aluminum being catalyzed by the corrosion products of other metals.	K	The corrosion of Al is catalyzed by Cu that forms by a displacement reaction on the Al surface. This is a well-understood phenomenon.

T4-7	L	Dissolution of fiberglass contributes products to the solid inventory in the pool, based upon the ICET test plan. Initial rankings changed due to following email, rankings indicated separately: Please find attached a table with my re-assessments of some originally "high" ratings of issues. On examining these issues again, it was evident that I did not fully recognize the impact of: 1. Time of exposure after the initial LOCA. 2. The amount of the specific material in the containment (e.g., zinc). My re-assessments also reflect the results of continuing radiolysis modeling, albeit with less-than-optimal codes that have been jerry-rigged from our reactor modeling work. Nevertheless, if the fields are as high as Bob indicates, the eventual production of nitric acid, in addition to the elevation of the redox potential, becomes a critical issue.	K	Has been demonstrated by the ICET.
T4-8	M	Leaching of fiberglass produces products that inhibit metal corrosion.	PK	Possible, but to my knowledge not firmly established.

| T4-9 | L | Zn passivation in galvanized steel leads to enhanced corrosion of iron. Initial rankings changed due to following email, rankings indicated separately: Please find attached a table with my re-assessments of some originally "high" ratings of issues. On examining these issues again, it was evident that I did not fully recognize the impact of:
1. Time of exposure after the initial LOCA.
2. The amount of the specific material in the containment (e.g., zinc).
My re-assessments also reflect the results of continuing radiolysis modeling, albeit with less-than-optimal codes that have been jerry-rigged from our reactor modeling work. Nevertheless, if the fields are as high as Bob indicates, the eventual production of nitric acid, in addition to the elevation of the redox potential, becomes a critical issue. | K | This has been demonstrated experimentally. |

| T4-10 | L | Zn corrosion products contribute to solids loading. Initial rankings changed due to following email, rankings indicated separately: Please find attached a table with my re-assessments of some originally "high" ratings of issues. On examining these issues again, it was evident that I did not fully recognize the impact of:
1. Time of exposure after the initial LOCA.
2. The amount of the specific material in the containment (e.g., zinc).
My re-assessments also reflect the results of continuing radiolysis modeling, albeit with less-than-optimal codes that have been jerry-rigged from our reactor modeling work. Nevertheless, if the fields are as high as Bob indicates, the eventual production of nitric acid, in addition to the elevation of the redox potential, becomes a critical issue. | K | Zn corrodes to form solid corrosion products that will contribute to the solids inventory in the containment pool. |

| T4-11 | L | ZnO or Zn(OH)$_2$ may precipitate with other species to form solid corrosion products.   Initial rankings changed due to following email, rankings indicated separately: Please find attached a table with my re-assessments of some originally "high" ratings of issues. On examining these issues again, it was evident that I did not fully recognize the impact of:
1. Time of exposure after the initial LOCA.
2. The amount of the specific material in the containment (e.g., zinc).
My re-assessments also reflect the results of continuing radiolysis modeling, albeit with less-than-optimal codes that have been jerry-rigged from our reactor modeling work. Nevertheless, if the fields are as high as Bob indicates, the eventual production of nitric acid, in addition to the elevation of the redox potential, becomes a critical issue. | PK | Co-precipitation has been demonstrated experimentally, although the phenomenon cannot be predicted from first principles or quantitatively. |

| T4-12 | L | Zinc oxide and hydroxide dissolves at pH > PZC to form zincate ion (ZnO_2^{2-}). Initial rankings changed due to following email, rankings indicated separately: Please find attached a table with my re-assessments of some originally "high" ratings of issues. On examining these issues again, it was evident that I did not fully recognize the impact of:
1. Time of exposure after the initial LOCA.
2. The amount of the specific material in the containment (e.g., zinc).
My re-assessments also reflect the results of continuing radiolysis modeling, albeit with less-than-optimal codes that have been jerry-rigged from our reactor modeling work. Nevertheless, if the fields are as high as Bob indicates, the eventual production of nitric acid, in addition to the elevation of the redox potential, becomes a critical issue. | K | The conditions under which dissolution occurs are well characterized experimentally and thermodynamically. |

T4-13	L	Destruction of Zn-based coatings will contribute to the inventory of corrosion products, including zinc species in the solution. Initial rankings changed due to following email, rankings indicated separately: Please find attached a table with my re-assessments of some originally "high" ratings of issues. On examining these issues again, it was evident that I did not fully recognize the impact of: 1. Time of exposure after the initial LOCA. 2. The amount of the specific material in the containment (e.g., zinc). My re-assessments also reflect the results of continuing radiolysis modeling, albeit with less-than-optimal codes that have been jerry-rigged from our reactor modeling work. Nevertheless, if the fields are as high as Bob indicates, the eventual production of nitric acid, in addition to the elevation of the redox potential, becomes a critical issue.	PK	While this true, the rate with which Zn dissolution products accumulate from coatings cannot be predicted quantitatively. The formation of zinc carbonates, hydroxides, and phosphates can be predicted thermodynamically.
T4-14		Do not understand the term "seal table")?		

T4-15	L	Fire barriers containing silicon-based materials might become a significant source of solid products. Initial rankings changed due to following email, rankings indicated separately: Please find attached a table with my re-assessments of some originally "high" ratings of issues. On examining these issues again, it was evident that I did not fully recognize the impact of: 1. Time of exposure after the initial LOCA. 2. The amount of the specific material in the containment (e.g., zinc). My re-assessments also reflect the results of continuing radiolysis modeling, albeit with less-than-optimal codes that have been jerry-rigged from our reactor modeling work. Nevertheless, if the fields are as high as Bob indicates, the eventual production of nitric acid, in addition to the elevation of the redox potential, becomes a critical issue.	PK	Dissolution phenomenon are characterized empirically, but insufficient data are available for predicting the accumulation of products.
T4-16	M	Organic material could increase the buoyancy of debris, thereby aiding its transport.	K	The floatation of minerals is a much understood phenomenon in the mining industry.
T4-17	H	Leaching of submerged coatings could contribute species to the containment pool, including Pb from Pb-based paints.	PK	The general principles of leaching are understood from the mining industry, but the rates cannot be predicted quantitatively at the current time.

T4-18	H	RCP oil is stored in tanks in containment. The failure of these tanks would contribute significantly to the inventory of foreign material in the containment pool.	PK	Since the oil is a liquid, it is difficult to predict exactly the consequences of the rupture of an oil storage tank.
T4-19	L	Biofilms that may form on metal surfaces could lead to the production of acids and hence increase metal corrosion rates.	PK	The time frame is too short for this to be a serious issue. Also, oxidizing bacteria, which might produce organic acids, would be inhibited if reducing conditions exist within containment. Finally, the buffer capacity of the coolant should overwhelm any organic acid formed.
T4-20	L	General corrosion enhanced under reducing conditions by SRBs (sulfate reducing bacteria).	PK	The time frame is too short for this to be a serious issue.
T4-21	M	SRBs might cause hydrogen embrittlement in submerged components.	PK	Occurs only on iron and carbon steels, but the time is too short for this to be a significant issue.
T4-22	M	Bacteria grow in pre-existing debris beds and contribute to screen clogging.	PK	Again, the time frame is too short for this to be a serious issue.
T4-23	M	Chlorides form via the radiolytic breakdown of cable insulation.	K	The rate of radiolytic degradation of many insulation materials is well-known and it is possible to estimate the chloride release rate if the ionizing radiation dose rate is known. Unless the dose rates are very high, the release rate is expected to be low and of little consequence compared with other sources. See T3-5.

T4-24	M	Oxide spalling off reactor fuel due to the high temperature gradient across the cladding wall. Initial rankings changed due to following email, rankings indicated separately: Please find attached a table with my re-assessments of some originally "high" ratings of issues. On examining these issues again, it was evident that I did not fully recognize the impact of: 1. Time of exposure after the initial LOCA. 2. The amount of the specific material in the containment (e.g., zinc). My re-assessments also reflect the results of continuing radiolysis modeling, albeit with less-than-optimal codes that have been jerry-rigged from our reactor modeling work. Nevertheless, if the fields are as high as Bob indicates, the eventual production of nitric acid, in addition to the elevation of the redox potential, becomes a critical issue.	K	This phenomenon can be predicted, in principle, from a thermal stress analysis. Sufficient spalling may compromise the integrity of the cladding, thereby increasing the risk of fission product release.
T4-25	H	Radio nuclides transported into the debris bed will alter the local chemical conditions.	K	See xxxx
T4-26	H	Radiolysis results in changes in oxidation states of materials and determines the reaction products formed.	K	See xxxx

T4-27	M	Radiolysis inh bits the formation of biofilms. Initial rankings changed due to following email, rankings indicated separately: Please find attached a table with my re-assessments of some originally "high" ratings of issues. On examining these issues again, it was evident that I did not fully recognize the impact of: 1. Time of exposure after the initial LOCA. 2. The amount of the specific material in the containment (e.g., zinc). My re-assessments also reflect the results of continuing radiolysis modeling, albeit with less-than-optimal codes that have been jerry-rigged from our reactor modeling work. Nevertheless, if the fields are as high as Bob indicates, the eventual production of nitric acid, in addition to the elevation of the redox potential, becomes a critical issue.	PK	This is true at high dose rates, but may not be so at low dose rates.
T4-28	H	Radiolysis causes changes in the redox potential in the debris bed and the containment pool.	K	This effect can be predicted using radiolysis models for water and the mixed potential model.
T4-29	H	Radiolysis of chloride-containing water produces HClO (hypochlorous acid), which enhances the corrosion rates of metals and alloys.	K	This effect can be predicted using radiolysis models for water and the mixed potential model.
T4-30	L	Radiolysis enhances the agglomeration of chemical species.	UK	Reviewer is unaware of this effect.

T4-31	L	Radiolysis may convert CO_2 into carbonates.	UK	Since the oxidation state of carbon in CO2 and CO32- is the same (+4), it is difficult to see how radiolysis can convert one into the other.
T4-32	H	Erosion-corrosion of pump components due to particulates in the water.	K	This is a well established phenomenon that could easily lead to degradation of pump performance.
T4-33	M	Pump seal degradation might contribute to debris inventory. Potential amount of debris is probably small compared with that from other sources.	K	This is a well established phenomenon that could easily lead to degradation of pump performance.
T4-34	L	Failure of steam generator may allow contamination of the containment pool with secondary side coolant. The chemistry of the secondary coolant is such that the impact would be minor and may even be beneficial.	K	The chemistry of the secondary coolant is well-defined and its impact on the containment pool chemistry is readily predicted.
T4-35	M	Precipitation of solids in the heat exchanger may reduce the efficiency of heat transfer. This is potentially important, in terms of heat removal from the core, but in the event of a LOCA the majority of the heat will be removed by coolant flowing from the break.	PK	Precipitation can be predicted qualitatively and possibly semi-quantitatively if the chemistry is well-defined. Recirculated containment pool water may make this prediction difficult.
T4-36	M	Drop in pH of evolving containment pool water may cause the precipitation of aluminum (and other metal) hydroxides. It is not clear what process could cause such a drop in the face of the high buffer capacity.	UK	The process causing the reduction in pH would need to be first identified.

T4-37	H	Precipitation of solids in the core could impede heat transfer from the fuel and lead to overheating of the fuel, ultimately resulting in fuel cladding failure and the release of fission products.	PK	Precipitation can be predicted qualitatively and possibly semi-quantitatively if the chemistry is well-defined. Recirculated containment pool water may make this prediction difficult.
T4-38	H	Spalling could lead to cladding failure and the release of fission products to containment.	K	The oxidation and spalling of zirconium alloys is well characterized and, if the chemistry of the environment can be defined, the phenomenon can be predicted.
T4-39	L	Amorphous silica forms on the surfaces of precipitates and makes less likely their transport to the sumo screen. I do not see the chemical basis for this postulate.	UK	The process responsible for the proposed effect would need to be first identified.
T4-40	M	Co-precipitation and ripening would produce larger precipitates that are less l kely to be transported to the screens. That may be correct, but the larger size might greatly increase the probability of entrapment of those particles that are transported to the screens.	PK	The effect of size on the tendency of particles to settle, be transported, and to be trapped can be predicted if the flow field can be defined to characterize the hydrodynamic forces involved.

| T4-41 | M | Turbulent flow, particularly under impingement conditions, may cause the rapid loss of material from a surface and hence contribute to the debris inventory in the pool.   Initial rankings changed due to following email, rankings indicated separately: Please find attached a table with my re-assessments of some originally "high" ratings of issues. On examining these issues again, it was evident that I did not fully recognize the impact of:
1. Time of exposure after the initial LOCA.
2. The amount of the specific material in the containment (e.g., zinc).
My re-assessments also reflect the results of continuing radiolysis modeling, albeit with less-than-optimal codes that have been jerry-rigged from our reactor modeling work. Nevertheless, if the fields are as high as Bob indicates, the eventual production of nitric acid, in addition to the elevation of the redox potential, becomes a critical issue. | K | Erosion-corrosion, scouring, and impingement attack are well defined and this phenomenon could, in principle, be defined with sufficient accuracy for the present purposes. |

APPENDIX I

PIRT Panelist Evaluation of T5 Phenomena:

ECCS Recirculation from 15 Days until 30 Days into

LOCA

Table I-1: ECCS Recirculation (15 days to 30 days)[a,b]

Issue Number	Phenomena	Description	Implications
T5-1	Organic radiolysis	Radiolysis of organics leads to carbonate formation	1. Create nucleation sites for species precipitation 2. scavenge dissolved Ca to impede formation of other compounds 3. Additional solid species contribution which leads to degraded system performance (head loss, etc.)

[a] Sump pool temperatures are expected to range between 40 – 60C during this time period.
[b] Phenomena listed in earlier time periods (i.e., T1, T2, T3, T4) may continue to be active or important during this time.

Table I-2: Apps T5 Ranking Summary

Issue Number	Importance (H, M, L)	Rationale	Knowledge (K, PK, UK)	Rationale
T5-1	L	Amount of organics likely to be small in relation to other debris and reactants.	UK	Further study required to quantify effects.

Table I-3: Chen T5 Ranking Summary

Issue Number	Importance (H, M, L)	Rationale	Knowledge (K, PK, UK)	Rationale
T5-1	No comment		No comment	

Table I-4: Delegard T5 Ranking Summary

Issue Number	Importance (H, M, L)	Rationale	Knowledge (K, PK, UK)	Rationale
T5-1	M	Radiolysis of organic materials likely to lead to other products before carbonate.	PK	Testing of radiolysis of organic materials has not been done for the post-LOCA system.

Table I-5: Litman T5 Ranking Summary

Issue Number	Importance (H, M, L)	Rationale	Knowledge (K, PK, UK)	Rationale
T5-1	L	Unless there is a significant introduction of organic materials after the initial event this will not be significant term.	PK	This is dependent upon ingress of oils (degraded pumps), or biologicals (raw water leaks).

Table I-6: MacDonald T5 Ranking Summary

Issue Number	Importance (H, M, L)	Rationale	Knowledge (K, PK, UK)	Rationale
T5-1	M	Radiolysis of organic material leads to carbonate formation and ultimately to the precipitation of calcium carbonate, which contributes to the solid inventory in the pool. This is possible, but whether carbonate forms will depend upon the redox potential. Initial rankings changed due to following email, rankings indicated separately: Please find attached a table with my re-assessments of some originally "high" ratings of issues. On examining these issues again, it was evident that I did not fully recognize the impact of: 1. Time of exposure after the initial LOCA. 2. The amount of the specific material in the containment (e.g., zinc). My re-assessments also reflect the results of continuing radiolysis modeling, albeit with less-than-optimal codes that have been jerry-rigged from our reactor modeling work. Nevertheless, if the fields are as high as Bob indicates, the eventual production of nitric acid, in addition to the elevation of the redox potential, becomes a critical issue.	PK	The radiolysis of water is fairly well-defined, but the redox conditions necessary for the formation of carbonate is not.

www.ingramcontent.com/pod-product-compliance
Lightning Source LLC
Chambersburg PA
CBHW081105170526
45165CB00008B/2337